SOLUTIONS MANUAL FOR PERSPECTIVES ON STRUCTURE AND MECHANISM IN ORGANIC CHEMISTRY

SOLUTIONS MANUAL FOR PERSPECTIVES ON STRUCTURE AND MECHANISM IN ORGANIC CHEMISTRY

Third Edition

Felix A. Carroll
Emeritus Professor
Davidson College, North Carolina, United States

WILEY

Published by John Wiley & Sons, Inc., Hoboken, New Jersey.
Published simultaneously in Canada.

Library of Congress Cataloging-in-Publication Data applied for
Paperback ISBN: 9781119808657; ePDF: 9781119808671; epub: 9781119808688

Cover Design by Wiley
Cover Image: © berCheck/Shutterstock; MOLEKUUL/SCIENCE PHOTO LIBRARY; ALFRED PASIEKA/SCIENCE PHOTO LIBRARY; Calicene image created by and courtesy of Felix A. Carroll

Set in 10/12pt PalatinoLTStd by Straive, Pondicherry, India

Contents

CHAPTER 1

Fundamental Models of Organic Chemistry

1.1. An answer to this question should be stated in terms of macroscopic phenomena. A historical exposition provides a rationale for the basis of contemporary chemistry, and several monographs on the history of chemistry can be used to summarize the ideas and observations that led to contemporary chemistry theory.[1,2,3]

1.2. See, for example, figure 2D in Tieu, P.; Yan, X.; Xu, M.; et al. *Small* **2021**, *17*, 2006482.

a. Transmission electron microscopy was used in this study.

b. The eye sees a macroscopic image on a computer screen or on a printed page.

For another example, see the image of kekulene reported by Pozo, I.; Majzik, Z.; Pavliček, N.; et al. *J. Am. Chem. Soc.* **2019**, *141*, 15488.

1.3. **a.** Two alternative geometries and their elimination on the basis of number of isomers are:

i. Square planar. There would be two isomers of CH_2Cl_2, one "cis," in which the Cl–C–Cl bond angle is 90°, and one "trans," in which the Cl–C–Cl bond angle is 180°.

ii. Square pyramid. Similarly, there would be two isomers of CH_2Cl_2 with the chlorines in the square base plus another isomer with one chlorine at the apex of the pyramid.

---

[1] Asimov, I. *A Short History of Chemistry*; Anchor Books: Garden City, NY, 1965.

[2] Ihde, A. J. *The Development of Modern Chemistry*; Harper & Row: New York, 1964.

[3] See, for example, Butterfield, H. *The Origins of Modern Science*, 1300-1800, Revised Edition; The Free Press: New York, 1965.

Solutions Manual for Perspectives on Structure and Mechanism in Organic Chemistry, Third Edition.
Felix A. Carroll.

b. In all answers, a substituent is presumed to occupy a position previously occupied by a hydrogen atom in the parent structure.[4]

i. If benzene had the structure we now call fulvene, there should be three different derivatives with the formula C_6H_5Cl.

ii. If benzene had the structure we now call Dewar benzene, there would be two and only two isomers with the formula C_6H_5Cl.

iii. If benzene had the structure we now call benzvalene, there would be three possible isomers with the formula C_6H_5Cl.

iv. If benzene had the structure we now call prismane, there would be only one isomer with the formula C_6H_5Cl, but there would be four isomers (including a pair of enantiomers) with the formula $C_6H_4Cl_2$.

v. If benzene had the structure we now call [3]radialene, there would be one and only one isomer with the formula C_6H_5Cl, but there would be four possible isomers (shown below) with the formula $C_6H_4Cl_2$.

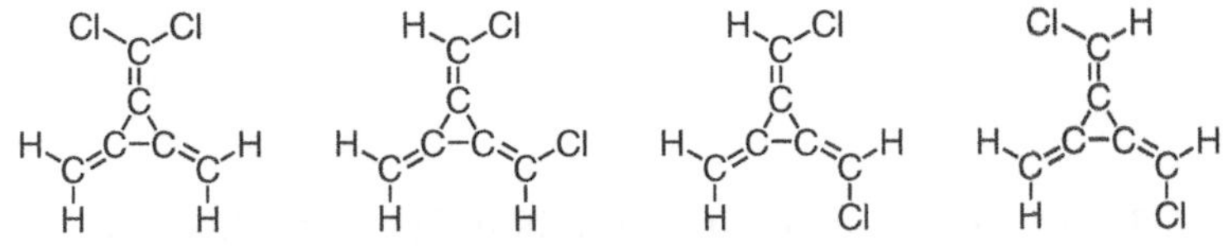

vi. There are also acyclic structures with the formula C_6H_6, such as 2,4-hexadiyne, and they may be analyzed similarly. For example, if benzene were 2,4-hexadiyne, then there would be one and only one C_6H_5Cl, but there could be only two structures with the formula $C_6H_4Cl_2$.

c. One can never know that something that has not been tested is like something else to which it seems similar. However, it seems unproductive to dwell on this possibility until there is an experimental result that could be rationalized with a structure for chloromethane that is different from the tetrahedral structure of methane. The spectroscopic results for chloromethane are consistent with a tetrahedral geometry.

1.4. The data and equations are given in Bondi, J. *J. Phys. Chem.* **1964**, *68*, 441.

For *n*-pentane,

$$V_W = 2\times13.67+3\times10.23 = 58.03\ \text{cm}^3/\text{mol}$$

$$A_W = 3\times1.35+2\times2.12 = 8.29\times10^9\ \text{cm}^2/\text{mol}$$

These results agree with those given by the general formulas for *n*-alkanes:

$$V_W = 6.88+10.23N_C = 6.88+10.23\times5 = 58.03\ \text{cm}^3/\text{mol}$$

$$A_W = 1.54+1.35N_C = 1.54+1.35\times5 = 8.29\times10^9\ \text{cm}^2/\text{mol}$$

[4] For a discussion of the number of isomers of benzene, see Reinecke, M. G. *J. Chem. Educ.* **1992**, 69, 859 and references therein.

For isopentane,

$$V_W = 3\times13.67+10.23+6.78 = 58.02\ \text{cm}^3/\text{mol}$$

$$A_W = 3\times2.12+1.35+0.57 = 8.28\times10^9\ \text{cm}^2/\text{mol}$$

For neopentane,

$$V_W = 4\times13.67+3.33 = 58.01\ \text{cm}^3/\text{mol}$$

$$A_W = 4\times2.12+0 = 8.48\times10^9\ \text{cm}^2/\text{mol}$$

1.5. Kiyobayashi, T.; Nagano, Y.; Sakiyama, M.; et al. *J. Am. Chem. Soc.* **1995**, *117*, 3270.

$$81.81+29.01 = 110.82\ \text{kcal/mol}.$$

1.6. Turner, R. B.; Goebel, P.; Mallon, B. J.; et al. *J. Am. Chem. Soc.* **1968**, *90*, 4315. Also see Hautala, R. R.; King, R. B.; Kutal, C. in Hautala, R. R.; King, R. B.; Kutal, C., Eds. *Solar Energy: Chemical Conversion and Storage*; Humana Press: Clifton, NJ, 1979; p. 333. The difference in heats of hydrogenation indicates that quadricyclane is less stable than norbornadiene by 24 kcal/mol, so this is the potential energy storage density for the photochemical reaction.

1.7. Pilcher, G.; Parchment, O. G.; Hillier, I. H.; et al. *J. Phys. Chem.* **1993**, *97*, 243.

$$\begin{array}{ll}
C_8H_{12}O_{2\,(s)} \rightarrow C_8H_{12}O_{2\,(g)} & \Delta H_s = 23.71\,\text{kcal/mol} \\
8CO_{2\,(g)} + 6H_2O_{\,(l)} \rightarrow C_8H_{12}O_{2\,(s)} + 10O_{2(g)} & -\Delta H_c = 1042.90\,\text{kcal/mol} \\
8C_{(graphite)} + 8O_{2\,(g)} \rightarrow 8CO_{2\,(g)} & \Delta H_f = 8(-94.05) = -752.4\ \text{kcal/mol} \\
6H_{2\,(g)} + 3O_{2\,(g)} \rightarrow 6H_2O_{\,(l)} & \Delta H_f = 6(-68.32) = -409.92\,\text{kcal/mol} \\
\hline
8C_{(graphite)} + 6H_{2(g)} + O_{2\,(g)} \rightarrow C_8H_{12}O_{2\,(g)} & \Delta H_f = -95.71\,\text{kcal/mol}
\end{array}$$

1.8. See Davis, H. E.; Allinger, N. L.; Rogers, D. W. *J. Org. Chem.* **1985**, *50*, 3601.

$$\begin{aligned}\Delta H_f(\text{phenylethyne}) &= \Delta H_f(\text{phenylethane}) - \Delta H_r(\text{phenylethyne}) \\ &= 7.15-(-66.12) = 73.27\ \text{kcal/mol}\end{aligned}$$

1.9. **a.** −632.6 ± 2.2 kJ/mol. Roux, M. V.; Temprado, M.; Jiménez, P.; et al. *J. Phys. Chem. A* **2006**, *110*, 12477.
b. 2-acetylthiophene is 4.7 kJ/mol more stable than 3-acetylthiophene in the gas phase. Roux, M. V.; Temprado, M.; Jiménez, P.; et al. *J. Phys. Chem. A* **2007**, *111*, 11084.

1.10. Wiberg, K. B.; Hao, S. *J. Org. Chem.* **1991**, *56*, 5108.

$$\begin{aligned}\Delta H_r(\textit{cis}\text{-3-methyl-2-pentene}) &= \Delta H_r(\text{2-ethyl-1-butene}) - \Delta\Delta H_f \\ &= -10.66-(-1.65) = -9.01\ \text{kcal/mol}\end{aligned}$$

1.11. Fang, W.; Rogers, D. W. *J. Org. Chem.* **1992**, *57*, 2294.

$$\begin{array}{ll} cis\text{-1,3,5-hexatriene} + 3H_2 \rightarrow n\text{-hexane} & \Delta H = -81.0\,\text{kcal/mol} \\ n\text{-hexane} \rightarrow \text{1,5-hexadiene} + 2H_2 & \Delta H = +60.3\,\text{kcal/mol} \\ \hline cis\text{-1,3,5-hexatriene} + H_2 \rightarrow \text{1,5-hexadiene} & \Delta H_r = -20.7\,\text{kcal/mol} \end{array}$$

$$\begin{array}{ll} trans\text{-1,3,5-hexatriene} + 3H_2 \rightarrow n\text{-hexane} & \Delta H = -80.0\,\text{kcal/mol} \\ n\text{-hexane} \rightarrow \text{1,5-hexadiene} + 2H_2 & \Delta H = +60.3\,\text{kcal/mol} \\ \hline trans\text{-1,3,5-hexatriene} + H_2 \rightarrow \text{1,5-hexadiene} & \Delta H_r = -19.7\,\text{kcal/mol} \end{array}$$

1.12. **a.** Using equation 1.9:

$$\begin{aligned} \Delta H_f &= 6(-146) + 16(-124.2) + 11(6.64) + 26(9.29) \\ &\quad + 5(10.2) + 7(231.3) + 16(52.1) \\ &= -44.92\,\text{kcal/mol} \end{aligned}$$

b. Using equation 1.12:

$$\begin{aligned} \Delta H_f &= -17.89 + 6(-2.15) + 5(-2.83) + 0(-7.74) + 0(-13.49) \\ &= -44.94\,\text{kcal/mol} \end{aligned}$$

1.13. See Smyth, C. P. in Weissberger, A.; Rossiter, B. W., Eds. *Physical Methods of Chemistry*, Vol. 1, Part 4; Wiley-Interscience: New York, 1972; pp. 397–429.

a. The gas phase dipole moments for CH_3–F, CH_3–Cl, CH_3–Br, and CH_3–I are 1.81, 1.87, 1.80 and 1.64 D, respectively. Using the bond length data in Table 1.1 and rewriting equation 1.23 lead to the following partial charges on F, Cl, Br, and I, respectively: −0.27, −0.22, −0.19, −0.16.

b. The dipole moments do not show a monotonic trend along the series because a dipole moment is a *product* of two terms. In the series of methyl halides, one term (the partial charge) goes down and the other term (bond length) goes up. The product of these two terms is a maximum at the second member of the series (X = Cl). Note that the assumption that only the carbon and halogen atoms are charged is an over-simplification. An Extended Hückel calculation indicates that the three methyl hydrogen atoms also bear some charge.

1.14. Because Pauling electronegativities are computed from the properties of atoms in molecules, they generally cannot be computed for the inert gases. However, krypton and xenon fluorides are known, and electronegativities of krypton and xenon were reported by Meek, T. L. *J. Chem. Educ.* **1995**, *72*, 17.

1.15. Using equation 1.47 leads to a value of 2.62 for λ_C^2. Therefore, the hybridization of carbon orbitals used for carbon–carbon bonds is $sp^{2.62}$. The relationship

$$2\left[\frac{1}{1+2.62}\right] + 2\left[\frac{1}{1+\lambda_H^2}\right] = 1$$

then gives a λ_C^2 value of 3.47 for the carbon orbitals used for the carbon–hydrogen bonds.

1.16. Mastryukov, V. S.; Schaefer, H. F., III.; Boggs, J. E. *Acc. Chem. Res.* **1994**, *27*, 242. Also see the discussion in Gilardi, R.; Maggini, M.; Eaton, P. E. *J. Am. Chem. Soc.* **1988**, *110*, 7232.

a. As the bond angle increases, the C–C bond length decreases. Conversely, as the bond angle decreases, the C–C bond length increases.

b. The larger the α, the greater the contribution of p character to the orbital of C2 used for the C2–C3 bond. This means greater s character in the orbital of C2 used for the C1–C2 bond, which results in a shorter C1–C2 bond. The same result can be rationalized using the VSEPR approach. As the angle α increases, there is less repulsion of the electrons comprising the C1–C2 bond with the electrons in the C2–C3 bond. This allows the electrons in the C1–C2 bond to move closer to C2, thus decreasing the bond length.

1.17. Maksić, Z. B.; Randić, M. *J. Am. Chem. Soc.* **1970**, *92*, 424. The bond lengths are a function of the hybridization of the carbon atoms.

a. ethyne, ethene, cyclopropane, cyclobutane, ethane.

b. 1,3-butadiyne, 1-butene-3-yne, 1,3-butadiene, propene, 2-methylpropene, 2-methylpropane, ethane.

1.18. a. According to the bent bond formulation, the electrons in the bent C–C bonds are pulled in toward the other olefinic carbon atom, so the electrons in these bonds repel the electrons in the carbon–hydrogen bonds less than they would in propane. Therefore, the H–C–H bond angle opens to a larger value.

b. The electrons in formaldehyde should be pulled even more strongly away from the carbon atom than is the case in ethene. Therefore, the repulsion of electrons in either C–O bond with electrons in a C–H bond is even less than the repulsion of electrons in the C–C bonds with electrons in a C–H bond in ethene. Therefore, the H–C–H bond angle in formaldehyde should be greater than that in ethene.

1.19. Based on an H–C–H angle of 116.2° for ethene, Robinson, E. A.; Gillespie, R. J. *J. Chem. Educ.* **1980**, *57*, 329 (appendix, p. 333) reported $sp^{2.26}$ or 30.6% s character for the C–H bond. Using 117° for the H–C–H angle[5] leads to $sp^{2.20}$, or 31.2% s character. For formaldehyde, using an H–C–H angle of 125.8° similarly leads to 36.9% s character for the carbon orbital used for carbon–hydrogen bonding.[6]

1.20. a. The formula is given by Newton, M. D.; Schulman, J. M.; Manus, M. M. *J. Am. Chem. Soc.* **1974**, *96*, 17. Rewrite equation 1.52 as $J = 5.7 \times (\%\,s) - 18.4\,\text{Hz}$. Then $500/(1+\lambda^2) = 5.7 \times (\%\,s) - 18.4$. Now let $\%\,s = 100/(1+\lambda^2)$ and solve for λ^2. It turns out to be just under 3. Thus, the equation is approximately correct for orbitals that are roughly sp^3-hybridized, but it is not exact for other orbitals.

b. The equation is

$$r_{\text{C–H}} = 1.1597 - (4.17 \times 10^{-4})(500)/(1+\lambda^2)$$

---

[5] (a) Bowen, H. J. M.; Donohue, J.; Jenkin, D. G.; et al., comps. *Tables of Interatomic Distances and Configuration in Molecules and Ions*, Special Publication No. 11; Chemical Society (London): Burlington House, W.1, London, 1958. (b) Supplement, 1965, p. M 78s.

[6] Reference 5(b), p. M 109.

so

$$r_{C-H} = 1.1597 - 0.209/(1+\lambda^2)$$

This equation is equivalent to

$$r_{C-H} = 1.1597 - 2.09\times 10^{-3}(\rho_{C-H})$$

where ρ_{C-H} is percent s character, which is defined as $100/(1+\lambda^2)$. This is the form of the equation given by Muller, N.; Pritchard, D. E. *J. Chem. Phys.* **1959**, *31*, 1471.

1.21. a. Here are calculations based on literature values for H–C–H bond angles and assuming that all molecules have planar carbon skeletons. (That is necessarily true only for cyclopropane.) Note that the calculated values depend on the choice of literature values for the bond angles.

	Cyclopropane[a]	Cyclobutane[b]	Cyclopentane[c]
∠H–C–H	118°	114°	109.5°
Using the formula $1+\lambda_i^2\cos\theta = 0$,			
λ_i^2:	2.13	2.459	2.996
Fraction s in C–H:	0.319	0.289	0.25
Fraction p in C–H:	0.681	0.711	0.75

[a] Reference 5(b), p. M98s.
[b] Reference 5(a), p. M 168.
[c] Reference 5(a), p. M 185.

Each carbon has 2 C–H bonds and 2 C–C bonds. Therefore, for a C–C bond of cyclopropane, the fractional s character is $0.5\times(1-2\times 0.319) = 0.181$.

Similarly,

	Cyclopropane	Cyclobutane	Cyclopentane
Fraction s in C–C:	0.181	0.211	0.25
Fraction p in C–C:	0.819	0.789	0.75
λ_j^2:	4.525	3.74	3.00
C–C–C interorbital angle:	102.77°	105.5°	109.47°

If the molecules are planar, then cyclopropane has $(102.77-60)/2 = 21.4°$ of angle strain at each carbon. Similarly, cyclobutane has 7.75° of angle strain, and cyclopentane has no angle strain.[7] As will be discussed in Chapter 4, cyclobutane and cyclopentane are not flat. The large fraction of p character in the cyclopropane carbon–carbon sigma bonds suggests that they might react to some extent like π bonds, which is the case. Note that the interorbital bond angle of cyclopropane is 102.77°, whereas the inter*nuclear* bond angle is required to be 60°. Thus, the cyclopropane bonds can be considered bent bonds.[8]

[7] This result for cyclopentane is based on the H–C–H bond angle reported in the literature. If the five carbon atoms of cyclopentane form a perfect pentagon, then the C–C–C bond angles are all 108°, so there is a slight amount of angle strain.

[8] Note also that cyclopropane has been described in terms of Walsh orbitals, which are based on p orbitals.

b. The acidity values can be correlated with *s* character by combining equations 1.52 and 1.53 to show a relationship between kinetic acidity and *s* character, and the results shown in Table 1.14 are consistent with such a relationship. By using the VSEPR concept, the very bent carbon–carbon bonds of cyclopropane (and to a lesser extent, cyclobutane) allow the electrons in the carbon–hydrogen bonds to be pulled closer to the carbon nucleus. That not only increases the H–C–H bond angle, but it also stabilizes a carbanion resulting from proton removal.

1.22. a. The predicted value, 110°, is very close to the value of 109.9° in Table 1.1.

b. As shown in the plot below, the error is smallest for H–C–X bond angles near 109.5° and becomes appreciable for bond angles 5° or 10° different from the normal tetrahedral value.

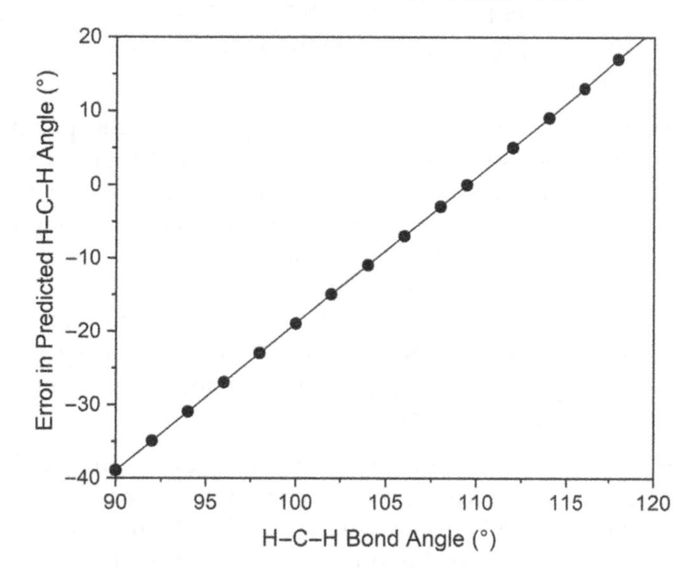

1.23. Plotting the data for ethane, ethene, and ethyne gives a linear ($R^2 = 0.999$) correlation:

$$\Delta H^\circ_{\text{acid}}\,(\text{kcal/mol}) = 461.9 - 53.96\,J_{^{13}\text{C–H}}$$

The gas phase acidity thus calculated for cyclopropane, 407.9 kcal/mol, is close to the experimental value, 410.7 kcal/mol. See Bartmess, J. E. in Mallard, W. G.; Linstrom, P. J., Eds. *NIST Webbook, NIST Standard Reference Database Number 69*; National Institute of Standards and Technology: Gaithersburg, MD (http://webbook.nist.gov). See also Fattahi, A.; McCarthy, R. E.; Ahmad, M. R.; et al. *J. Am. Chem. Soc.* **2003**, *125*, 11746.

1.24. Kass S. R.; Chou, P. K. *J. Am. Chem. Soc.* **1988**, *110*, 7899. $1 + \lambda^2 = 500/202$, so $\lambda^2 = 1.475$. Therefore, the percent *s* character is $100/2.475 = 40.4\%$. This is less than the 50% *s* character in acetylene C–H bonds, so acetylene should be more acidic.

1.25. If hybridization does not exist, then a quantification of hybridization is an artifice. Even though λ cannot be observed directly, however, it provides a more satisfying basis for correlating coupling constants, acidities, and bond angles with each other than would a purely empirical correlation of any two of these observables.

1.26. Zavitsas, A. A.; Matsunaga, N.; Rogers, D. W. *J. Phys. Chem. A* **2008**, *112*, 5734. The method cannot be applied to methane because a value for methyl hydrogens was not included in its derivation (and there is no need to include a parameter that applies to only one compound). For the other compounds, however, the R^2 for a correlation of literature and predicted values is 0.999. As the authors point out, "terms are not needed to account for repulsive or attractive 1,3 interactions, hyperconjugation, or for protobranching, rendering them irrelevant." This example reminds us that chemists should not be too quick to associate correlation with causation.

1.27. There is no literature reference for this problem, nor is there a single right answer. One response is that the concept of hybridization provides a useful conceptual model for understanding the bonding of carbon compounds without the need for carrying out molecular orbital calculations in which hybridized orbitals are not assumed. Therefore, hybridized orbitals are useful concepts but not physical attributes of atoms.

1.28. The answer to this question depends on the perspective of the responder. Organic chemists use pictorial representations because they can convey chemical meaning more compactly than mathematical expressions. Organic chemistry may become more mathematical as the role of computation becomes ever more important, but it is likely that the results of a mathematical analysis will continue to be interpreted in a largely pictorial form by the organic chemist. (Some readers may wish to speculate how this conclusion could change if artificial intelligence becomes a routine tool in organic chemistry.)

1.29. Coulson stated the basic paradox of chemistry. We live in a macroscopic world, but we explain that world in terms of unseen particles and unseen forces. To the chemist, atoms, bonds, and molecules are real and can be demonstrated. However, Coulson's reminder that these concepts are intangible reinforces the view that chemistry is based on models that are subject to revision if better models become available.

1.30. Hoffmann is correct in saying that psychology plays an important role in the acceptance of theories. However, scientists who truly understand the limits of human knowledge may be less tempted to make decisions on the basis of a strong conviction of causality than are nonscientists.

1.31. Weisberg's comments are consistent with the primary argument of this chapter only if there are multiple, complementary models for the question at hand. If there is only one model for a particular situation, then the relationship between precision and generality is less certain.

1.32. There is not a single correct answer to this question. One definition of compatibility is that two or more things can coexist without conflict. The question in this context is "without conflict with what?" It might be argued that two models that appear to be in conflict with each other conceptually could be considered compatible if they correctly predict the results of all related physical measurements. That is, they are compatible in their opacity if no test can show that one is correct and the other is not.

1.33. Each reader will have a different response to this question. Discussions may refer to the concept of variable hybridization, the different meanings of electronegativity, the underpinnings of valence bond theory and molecular orbital theory, the nature of a carbon–carbon double bond, and the importance of considering any chemical question in terms of more than one conceptual model.

CHAPTER 2

Introduction to Computational Chemistry

2.1. Substitute $(\alpha-\beta)$ for E in equations 2.15 and 2.16 and then solve for c_1 and c_2.

2.2. The π bond orders are the same for all three species because the only bonding between C1 and C2 and between C2 and C3 results from population of ψ_1. Therefore, $P_{12} = P_{23} = 2 \times 0.500 \times 0.707 = 0.707$. Similarly, the free valence indices for the cation, radical, and anion are all the same for each species.

For C1 and C3, $\mathcal{F}_1 = \mathcal{F}_3 = 4.732 - (3 + 0.707) = 1.025$.

For C2, $\mathcal{F}_2 = 4.732 - (3 + 0.707 + 0.707) = 0.318$.

2.3. The values must be the same for odd-numbered cyclic systems (such as cyclopropenyl, cyclopentadienyl, and so on) in which all atoms are equivalent by symmetry.

2.4. The secular determinants are shown below in the solutions for problem 2.5. Hückel molecular orbital calculations for many compounds of interest were reported in Heilbronner, E.; Straub, W. *Hückel Molecular Orbitals: HMO*; Springer-Verlag: New York, 1966.

Solutions Manual for Perspectives on Structure and Mechanism in Organic Chemistry, Third Edition.
Felix A. Carroll.

2.5. a. Hückel matrix for cyclobutadiene

$$\begin{vmatrix} X & 1 & 0 & 1 \\ 1 & X & 1 & 0 \\ 0 & 1 & X & 1 \\ 1 & 0 & 1 & X \end{vmatrix} = 0$$

$$\psi_4 = -0.500\phi_1 + 0.500\phi_2 - 0.500\phi_3 + 0.500\phi_4 \quad E = \alpha - 2.000\beta$$

$$\psi_3 = +0.500\phi_1 - 0.500\phi_2 - 0.500\phi_3 + 0.500\phi_4 \quad E = \alpha$$

$$\psi_2 = -0.500\phi_1 - 0.500\phi_2 + 0.500\phi_3 + 0.500\phi_4 \quad E = \alpha$$

$$\psi_1 = +0.500\phi_1 + 0.500\phi_2 + 0.500\phi_3 + 0.500\phi_4 \quad E = \alpha + 2.000\beta$$

Note: reversing all signs in any wave function does not change the energy or bonding relationships. Thus, we may also write

$$\psi_2 = +0.500\phi_1 + 0.500\phi_2 - 0.500\phi_3 - 0.500\phi_4$$

With 2, 1, 1, 0 electrons in the four MOs (populated from ψ_1 to ψ_4), the π energy is

$$E_\pi = 4\alpha + 4.000\beta$$

Electron densities:

$$\rho_1 = 1.000; \rho_2 = 1.000; \rho_3 = 1.000; \rho_4 = 1.000$$

Partial bond orders:

$$P_{12} = P_{23} = P_{34} = P_{41} = 0.5000$$

Free valence indices:

$$\mathcal{F}_1 = 0.732;\ \mathcal{F}_2 = 0.732;\ \mathcal{F}_3 = 0.732;\ \mathcal{F}_4 = 0.732$$

b. Hückel matrix for methylenecyclopropene

$$\begin{vmatrix} X & 1 & 0 & 0 \\ 1 & X & 1 & 1 \\ 0 & 1 & X & 1 \\ 0 & 1 & 1 & X \end{vmatrix} = 0$$

$$\psi_4 = +0.506\phi_1 - 0.749\phi_2 + 0.302\phi_3 + 0.302\phi_4 \quad E = \alpha - 1.481\beta$$

$$\psi_3 = +0.000\phi_1 - 0.000\phi_2 - 0.707\phi_3 + 0.707\phi_4 \quad E = \alpha - 1.000\beta$$

$$\psi_2 = -0.815\phi_1 - 0.254\phi_2 + 0.368\phi_3 + 0.368\phi_4 \quad E = \alpha + 0.311\beta$$

$$\psi_1 = +0.282\phi_1 + 0.612\phi_2 + 0.523\phi_3 + 0.523\phi_4 \quad E = \alpha + 2.170\beta$$

With 2, 2, 0, and 0 electrons in the four MOs, $E_\pi = 4\alpha + 4.962\beta$

Electron densities:

$$\rho_1 = 1.488;\ \rho_2 = 0.877;\ \rho_3 = \rho_4 = 0.818$$

Partial bond orders:

$$P_{12} = 0.758;\ P_{23} = P_{24} = 0.453;\ P_{34} = 0.818$$

Free valence indices:

$$\mathcal{F}_1 = 0.974;\ \mathcal{F}_2 = 0.068;\ \mathcal{F}_3 = 0.462;\ \mathcal{F}_4 = 0.462$$

c. Hückel matrix for fulvene

$$\begin{vmatrix} X & 1 & 0 & 0 & 1 & 1 \\ 1 & X & 1 & 0 & 0 & 0 \\ 0 & 1 & X & 1 & 0 & 0 \\ 0 & 0 & 1 & X & 1 & 0 \\ 1 & 0 & 0 & 1 & X & 0 \\ 1 & 0 & 0 & 0 & 0 & X \end{vmatrix} = 0$$

The resulting MOs and energy levels are shown in compact form in the following table, where the coefficients of each ϕ in ψ_1 are listed vertically under the heading "MO 1." Thus,

$$\psi_1 = 0.523\phi_1 + 0.429\phi_2 + 0.385\phi_3 + 0.385\phi_4 + 0.429\phi_5 + 0.247\phi_6$$

and the energy of ψ_1 is $\alpha + 2.115\beta$

	MO 1	MO 2	MO 3	MO 4	MO 5	MO 6
c\E	2.115	1.000	0.618	-0.254	-1.618	-1.861
1	0.523	-0.500	0.000	0.190	0.000	-0.664
2	0.429	0.000	-0.602	0.351	-0.372	0.439
3	0.385	0.500	-0.372	-0.280	0.602	-0.153
4	0.385	0.500	0.372	-0.280	-0.602	-0.153
5	0.429	0.000	0.602	0.351	0.372	0.439
6	0.247	-0.500	0.000	-0.749	0.000	0.357

With 2, 2, 2, 0, 0, and 0 electrons in the 6 MOs,

$$E_\pi = 6\alpha + 7.466\beta$$

Values of ρ_i are indicated on the diagonal (in bold) and values of P_{ij} are given by the off-diagonal elements in the following table:

	1	2	3	4	5	6
1	**1.0470**	.4491			.4491	.7586
2	.4491	**1.0923**	.7779			
3		.7779	**1.0730**	.5202		
4			.5202	**1.0730**	.7779	
5	.4491			.7779	**1.0923**	
6	.7586					**.6223**

For example, $\rho_1 = 1.0470$ and $P_{15} = 0.4491$. The free valence indices are

$$\mathcal{F}_1 = \mathcal{F}_5 = 0.075;\ \mathcal{F}_2 = 0.505;\ \mathcal{F}_3 = \mathcal{F}_4 = 0.434;\ \mathcal{F}_6 = 0.973$$

d. Hückel matrix for styrene

$$\begin{vmatrix} X & 1 & 0 & 0 & 0 & 1 & 1 & 0 \\ 1 & X & 1 & 0 & 0 & 0 & 0 & 0 \\ 0 & 1 & X & 1 & 0 & 0 & 0 & 0 \\ 0 & 0 & 1 & X & 1 & 0 & 0 & 0 \\ 0 & 0 & 0 & 1 & X & 1 & 0 & 0 \\ 1 & 0 & 0 & 0 & 1 & X & 0 & 0 \\ 1 & 0 & 0 & 0 & 0 & 0 & X & 1 \\ 0 & 0 & 0 & 0 & 0 & 0 & 1 & X \end{vmatrix} = 0$$

Table of molecular orbitals and energy levels:

	MO 1	MO 2	MO 3	MO 4	MO 5	MO 6	MO 7	MO 8
c\\E	2.136	1.414	1.000	.662	-.662	-1.000	-1.414	-2.136
1	.513	-.354	.000	-.334	-.334	.000	-.354	.513
2	.394	.000	-.500	-.308	.308	-.500	.000	-.394
3	.329	.354	-.500	.130	.130	.500	.354	.329
4	.308	.500	.000	.394	-.394	.000	-.500	-.308
5	.329	.354	.500	.130	.130	-.500	.354	.329
6	.394	.000	.500	-.308	.308	.500	.000	-.394
7	.308	-.500	.000	.394	-.394	.000	.500	-.308
8	.144	-.354	.000	.595	.595	.000	-.354	.144

With 2, 2, 2, 2, 0, 0, 0, and 0 in the 8 MOs,

$$E_\pi = 8\alpha + 10.424\beta$$

Values of ρ_i (in bold) and P_{ij} are summarized in the following table:

	1	2	3	4	5	6	7	8
1	**1.0000**	.6101				.6101	.4059	
2	.6101	**1.0000**	.6787					
3		.6787	**1.0000**	.6586				
4			.6586	**1.0000**	.6586			
5				.6586	**1.0000**	.6787		
6	.6101				.6787	**1.0000**		
7	.4059						**1.0000**	.9113
8							.9113	**1.0000**

Free valence indices:

$$\mathcal{F}_1 = 0.106;\ \mathcal{F}_2 = 0.443;\ \mathcal{F}_3 = 0.395;\ \mathcal{F}_4 = 0.415;$$

$$\mathcal{F}_5 = 0.395;\ \mathcal{F}_6 = 0.443;\ \mathcal{F}_7 = 0.415;\ \mathcal{F}_8 = 0.821$$

e. Hückel matrix for naphthalene

$$\begin{vmatrix} X & 1 & 0 & 0 & 0 & 0 & 0 & 0 & 1 & 0 \\ 1 & X & 1 & 0 & 0 & 0 & 0 & 0 & 0 & 0 \\ 0 & 1 & X & 1 & 0 & 0 & 0 & 0 & 0 & 0 \\ 0 & 0 & 1 & X & 0 & 0 & 0 & 0 & 0 & 1 \\ 0 & 0 & 0 & 0 & X & 1 & 0 & 0 & 0 & 1 \\ 0 & 0 & 0 & 0 & 1 & X & 1 & 0 & 0 & 0 \\ 0 & 0 & 0 & 0 & 0 & 1 & X & 1 & 0 & 0 \\ 0 & 0 & 0 & 0 & 0 & 0 & 1 & X & 1 & 0 \\ 1 & 0 & 0 & 0 & 0 & 0 & 0 & 1 & X & 1 \\ 0 & 0 & 0 & 1 & 1 & 0 & 0 & 0 & 1 & X \end{vmatrix} = 0$$

Table of molecular orbitals and energy levels:

	MO 1	MO 2	MO 3	MO 4	MO 5	MO 6	MO 7	MO 8	MO 9	MO 10
c\E	2.303	1.618	1.303	1.000	.618	-.618	-1.000	-1.303	-1.618	-2.303
1	.301	-.263	.400	.000	-.425	-.425	.000	.400	.263	.301
2	.231	-.425	.174	.408	-.263	.263	.408	-.174	-.425	-.231
3	.231	-.425	-.174	.408	.263	.263	-.408	-.174	.425	.231
4	.301	-.263	-.400	.000	.425	-.425	.000	.400	-.263	-.301
5	.301	.263	-.400	.000	-.425	.425	.000	.400	.263	-.301
6	.231	.425	-.174	.408	-.263	-.263	-.408	-.174	-.425	.231
7	.231	.425	.174	.408	.263	-.263	.408	-.174	.425	-.231
8	.301	.263	.400	.000	.425	.425	.000	.400	-.263	.301
9	.461	.000	.347	-.408	.000	.000	-.408	-.347	.000	-.461
10	.461	.000	-.347	-.408	.000	.000	.408	-.347	.000	.461

With 2, 2, 2, 2, 2, 0, 0, 0, 0, and 0 electrons in the ten MOs,

$$E_\pi = 10\alpha + 13.683\beta$$

Values of ρ_i (in bold) and P_{ij} are summarized in the following table:

	1	2	3	4	5	6	7	8	9	10
1	**1.0000**	.7246							.5547	
2	.7246	**1.0000**	.6032							
3		.6032	**1.0000**	.7246						
4			.7246	**1.0000**						.5547
5					**1.0000**	.7246				.5547
6					.7246	**1.0000**	.6032			
7						.6032	**1.0000**	.7246		
8							.7246	**1.0000**	.5547	
9	.5547							.5547	**1.0000**	.5182
10				.5547	.5547				.5182	**1.0000**

Free valence indices:

$\mathcal{F}_1 = \mathcal{F}_4 = \mathcal{F}_5 = \mathcal{F}_8 = 0.453$; $\mathcal{F}_2 = \mathcal{F}_3 = \mathcal{F}_6 = \mathcal{F}_7 = 0.404$;
$\mathcal{F}_9 = \mathcal{F}_{10} = 0.104$

f. Hückel matrix for biphenylene

$$\begin{vmatrix} X & 1 & 0 & 0 & 0 & 1 & 0 & 0 & 0 & 0 & 0 & 1 \\ 1 & X & 1 & 0 & 0 & 0 & 0 & 0 & 0 & 0 & 0 & 0 \\ 0 & 1 & X & 1 & 0 & 0 & 0 & 0 & 0 & 0 & 0 & 0 \\ 0 & 0 & 1 & X & 1 & 0 & 0 & 0 & 0 & 0 & 0 & 0 \\ 0 & 0 & 0 & 1 & X & 1 & 0 & 0 & 0 & 0 & 0 & 0 \\ 1 & 0 & 0 & 0 & 1 & X & 1 & 0 & 0 & 0 & 0 & 0 \\ 0 & 0 & 0 & 0 & 0 & 1 & X & 1 & 0 & 0 & 0 & 1 \\ 0 & 0 & 0 & 0 & 0 & 0 & 1 & X & 1 & 0 & 0 & 0 \\ 0 & 0 & 0 & 0 & 0 & 0 & 0 & 1 & X & 1 & 0 & 0 \\ 0 & 0 & 0 & 0 & 0 & 0 & 0 & 0 & 1 & X & 1 & 0 \\ 0 & 0 & 0 & 0 & 0 & 0 & 0 & 0 & 0 & 1 & X & 1 \\ 1 & 0 & 0 & 0 & 0 & 0 & 1 & 0 & 0 & 0 & 1 & X \end{vmatrix} = 0$$

Table of molecular orbitals and energy levels:

	MO 1	MO 2	MO 3	MO 4	MO 5	MO 6	MO 7	MO 8	MO 9	MO 10
c\E	2.532	1.802	1.347	1.247	.879	.445	-.445	-.879	-1.247	-1.347
1	.422	.164	-.225	.296	.147	-.368	.368	.147	-.296	.225
2	.225	.296	.147	.368	.422	-.164	-.164	-.422	.368	.147
3	.147	.368	.422	.164	.225	.296	-.296	.225	-.164	-.422
4	.147	.368	.422	-.164	-.225	.296	.296	.225	-.164	.422
5	.225	.296	.147	-.368	-.422	-.164	.164	-.422	.368	-.147
6	.422	.164	-.225	-.296	-.147	-.368	-.368	.147	-.296	-.225
7	.422	-.164	-.225	-.296	.147	.368	-.368	.147	.296	.225
8	.225	-.296	.147	-.368	.422	.164	.164	-.422	-.368	.147
9	.147	-.368	.422	-.164	.225	-.296	.296	.225	.164	-.422
10	.147	-.368	.422	.164	-.225	-.296	-.296	.225	.164	.422
11	.225	-.296	.147	.368	-.422	.164	-.164	-.422	-.368	-.147
12	.422	-.164	-.225	.296	-.147	.368	.368	.147	.296	-.225

	MO 11	MO 12
	-1.802	-2.532
1	.164	-.422
2	-.296	.225
3	.368	-.147
4	-.368	.147
5	.296	-.225
6	-.164	.422
7	-.164	-.422
8	.296	.225
9	-.368	-.147
10	.368	.147
11	-.296	-.225
12	.164	.422

With 2, 2, 2, 2, 2, 2, 0, 0, 0, 0, 0, and 0 electrons in the 12 MOs,

$$E_\pi = 12\alpha + 16.505\beta$$

Values of ρ_i (in bold) and P_{ij} are summarized in the following table:

	1	2	3	4	5	6	7	8	9	10	11	12
1	**1.0000**	.6830				.5648						.2634
2	.6830	**1.0000**	.6208									
3		.6208	**1.0000**	.6907								
4			.6907	**1.0000**								
5				.6208	.6208							
6	.5648				**1.0000**	.6830						
7					.6830	**1.0000**	.2634				.5648	
8						.2634	**1.0000**	.6830				
9							.6830	**1.0000**	.6208			
10								.6208	**1.0000**	.6907	.6208	
11									.6907	**1.0000**	**1.0000**	.6830
12	.2634						.5648			.6208	.6830	**1.0000**

Free valence indices:

$\mathcal{F}_1=\mathcal{F}_6=\mathcal{F}_7=\mathcal{F}_{12}=0.221; \mathcal{F}_2=\mathcal{F}_5=\mathcal{F}_8=\mathcal{F}_{11}=0.428; \mathcal{F}_3=\mathcal{F}_4=\mathcal{F}_9=\mathcal{F}_{10}=0.420.$

2.6. These parameters were included in the answers to question 2.5.

2.7. The energy levels are shown below. There is an orbital at $E=\alpha+2.414\beta$ for all of the radialenes and an orbital at $E=\alpha-2.414\beta$ for the even-numbered rings. See Fenet-Buchholz, J.; Boese, R.; Haumann, T.; et al. in Rappoport, Z., Ed. *The Chemistry of Dienes and Polyenes*, Vol. 1; John Wiley & Sons: New York, 1997, p. 25; Kutzelnigg, W. *J. Comp. Chem.* **2007**, *28*, 25.

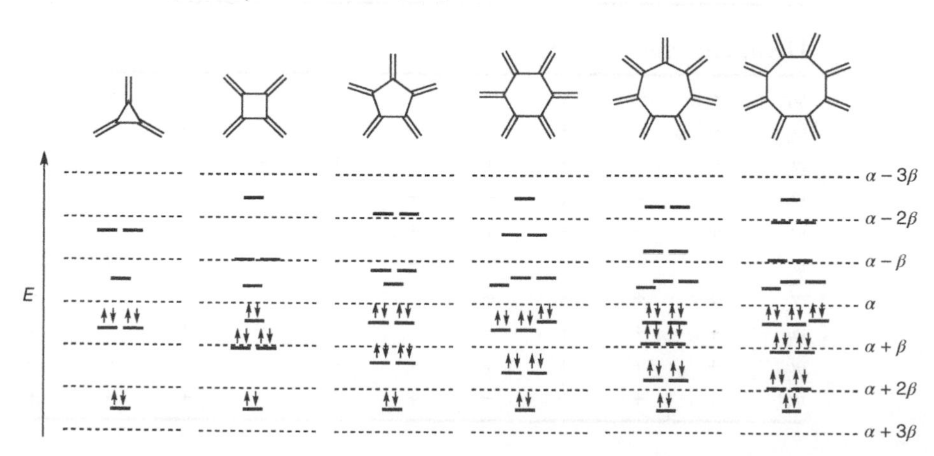

2.8. The orbital symmetries are indicated schematically below. A plus sign means that the coefficient is positive for a given atomic orbital, while a minus sign means that the coefficient is negative. A zero indicates a node at a particular position. The exact placement of the nodes may vary, but the important factors are (i) the proper number of nodes, (ii) the symmetry of the placement of the nodes, and (iii) increasing energy as one goes from ψ_n to ψ_{n+1}.

Heptatrienyl

ψ_7	+	−	+	−	+	−	+	
ψ_6	+	−	+	0	−	+	−	
ψ_5	+	−	−	+	−	−	+	
ψ_4	+	0	−	0	+	0	−	(NBMO)
ψ_3	+	+	−	−	−	+	+	
ψ_2	+	+	+	0	−	−	−	
ψ_1	+	+	+	+	+	+	+	

Octatetraene

ψ_8	+	−	+	−	+	−	+	−
ψ_7	+	−	+	−	−	+	−	+
ψ_6	+	−	0	+	−	0	+	−
ψ_5	−	+	+	−	−	+	+	−
ψ_4	+	+	−	−	+	+	−	−
ψ_3	+	+	0	−	−	0	+	+
ψ_2	+	+	+	+	−	−	−	−
ψ_1	+	+	+	+	+	+	+	+

Nonatetraenyl

ψ_9	+	−	+	−	+	−	+	−	+	
ψ_8	+	−	+	−	0	+	−	+	−	
ψ_7	+	−	+	+	−	+	+	−	+	
ψ_6	+	−	−	+	0	−	+	+	−	
ψ_5	+	0	−	0	+	0	−	0	−	(NBMO)
ψ_4	+	+	−	−	0	+	+	−	−	
ψ_3	+	+	+	−	−	−	+	+	+	
ψ_2	+	+	+	+	0	−	−	−	−	
ψ_1	+	+	+	+	+	+	+	+	+	

Decapentaene

ψ_{10}	+	−	+	−	+	−	+	−	+	−
ψ_9	+	−	+	−	+	+	−	+	−	+
ψ_8	+	−	+	+	−	+	−	−	+	−
ψ_7	−	+	+	−	+	+	−	+	+	−
ψ_6	+	−	−	+	+	−	−	+	+	−
ψ_5	+	+	−	−	+	+	−	−	+	+
ψ_4	+	+	−	−	−	+	+	+	−	−
ψ_3	+	+	+	−	−	−	−	+	+	+
ψ_2	+	+	+	+	+	−	−	−	−	−
ψ_1	+	+	+	+	+	+	+	+	+	+

2.9. Structures (a) and (c) are nonalternant. Structure (b) is odd alternant.

2.10. Write the equations for the sum of the NBMO coefficients of the atomic orbitals directly linked to a nonstarred atom.

$C_1 + C_3 + C_5 = 0$

$C_3 + C_5 = 0$

Because $(c_1 + c_3 + c_5) = (c_3 + c_5) = 0$, c_1 must equal 0. It may be surprising that the radical character on carbon 1 is 0 because we are calculating properties of a Lewis structure indicating a full radical at that position. This result is somewhat easier to understand by noting that resonance theory would predict the two resonance structures on the left below to be much more stable than the two antiaromatic resonance structures on the right.

Major contributors to resonance hybrid

Very minor contributors to resonance hybrid

This result is consistent with an HMO calculation:

Input Hückel matrix:

$$\begin{bmatrix} X & 1 & 0 & 0 & 0 \\ 1 & X & 1 & 0 & 0 \\ 0 & 1 & X & 1 & 0 \\ 0 & 0 & 1 & X & 1 \\ 0 & 1 & 0 & 1 & X \end{bmatrix} = 0$$

Table of molecular orbitals and energy levels:

	MO 1	MO 2	MO 3	MO 4	MO 5
c\E	2.136	.662	.000	-.662	-2.136
1	.261	-.657	.000	-.657	.261
2	.557	-.435	.000	.435	-.557
3	.465	.185	-.707	.185	.465
4	.435	.557	.000	-.557	-.435
5	.465	.185	.707	.185	.465

With 2, 2, 1, 0, and 0 electrons in the 5 MOs,

$$E_\pi = 5\alpha + 5.596\,\beta$$

Values of ρ_i (in bold) and P_{ij} are summarized in the following table:

	1	2	3	4	5
1	**1.0000**	.8629			
2	.8629	**1.0000**	.3574		.3574
3		.3574	**1.0000**	.6101	
4			.6101	**1.0000**	.6101
5		.3574		.6101	**1.0000**

Free valence indices:

$$\mathcal{F}_1 = 0.869;\ \mathcal{F}_2 = 0.154;\ \mathcal{F}_3 = 0.764;\ \mathcal{F}_4 = 0.512;\ \mathcal{F}_5 = 0.764$$

2.11. Methylenecyclopropene was reported by Billups, W. E.; Lin, L.-J.; Casserly, E. W. *J. Am. Chem. Soc.* **1984**, *106*, 3698 and by Staley, S. W.; Norden, T. D. *J. Am. Chem. Soc.* **1984**, *106*, 3699. For a discussion of the structure of methylenecyclopropene (dipole moment of 1.90 D) and a theoretical description that is more complex than the simple resonance description, see Norden, T. D.; Staley, S. W.; Taylor, W. H.; et al. *J. Am. Chem. Soc.* **1986**, *108*, 7912. Also see Bachrach, S. M. *J. Org. Chem.* 1990, *55*, 4961. In contrast to naphthalene, methylenecyclopropene was found to have nonzero q_i values. The results agree with the statement in the chapter about the nature of alternant and nonalternant systems. Resonance theory would have predicted a resonance hybrid with contribution from resonance structures having an aromatic cyclopropenyl cation bonded to a methylene anion substituent, so the resonance hybrid would be quite polar.

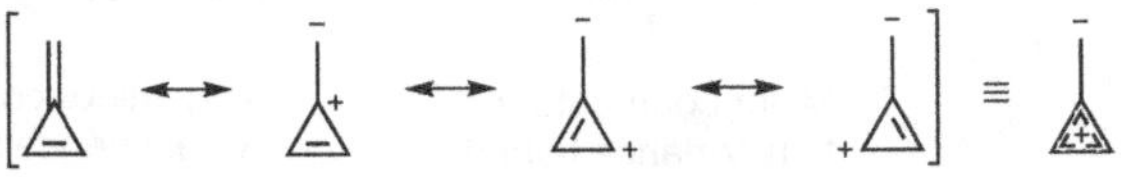

2.12. The argument is just the same as that for methylenecyclopropene: the stability of the aromatic cyclopropenyl cation makes the polar form more stable. In addition, the greater electronegativity of oxygen than carbon also enhances the polarity of the molecule. For experimental data, see Staley, S. W.; Norden, T. D.; Taylor, W. H.; Harmony, M. D. *J. Am. Chem. Soc.* **1987**, *109*, 7641; also see Bachrach, S. M. *J. Org. Chem.* **1990**, *55*, 4961;Wang, Y.; Fernández, I.; Duvall, M.; et al. *J. Org. Chem.* **2010**, *74*, 8252.

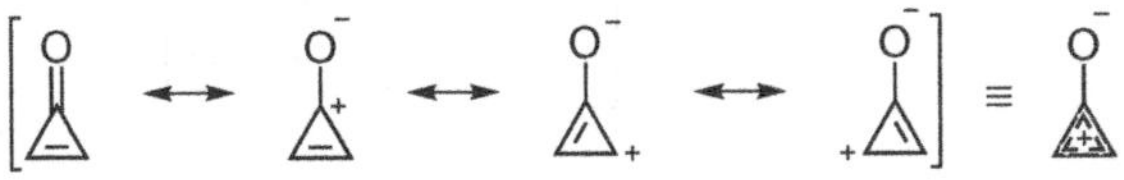

2.13. Overlap between carbons 1 and 4 and between carbons 2 and 3 decreases as the structure is stretched as shown in the problem, but the total energy of the structure stays the same. As just *one* side is stretched, however, the π energy approaches that of 1,3-butadiene. Therefore, the HMO delocalization energy for cyclobutadiene should be greater for a trapezoidal structure than for a rectangular structure. Of course, the HMO result ignores geometric restrictions of the σ framework.

2.14. Podlogar, B. L.; Glauser, W. A.; Rodriguez, W. R.; et al. *J. Org. Chem.* **1988**, *53*, 2127; Nakayama, M.; Ishikawa, H.; Nakano, T.; et al. *THEOCHEM* **1989**, *53*, 369; Sekiguchi, A.; Ebata, K.; Kabuto, C.; et al. *J. Am. Chem. Soc.* **1991**, *113*, 7081. The dianion is antiaromatic, having a triplet ground state with a preference for nonplanarity of the carbon skeleton.

2.15. HMO calculation for rotation about allyl radical:

a. Input Hückel matrix for starting allyl radical ($H_{2,3} = 1.00$)

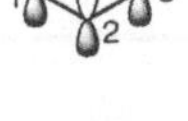

$$\begin{bmatrix} X & 1 & 0 \\ 1 & X & 1 \\ 0 & 1 & X \end{bmatrix} = 0$$

Table of molecular orbitals and energy levels:

	MO 1	MO 2	MO 3
c\E	1.414	.000	-1.414
1	.500	-.707	.500
2	.707	.000	-.707
3	.500	.707	.500

With 2, 1, and 0 electrons in the 3 MOs,

$$E_{\pi} = 3\alpha + 2.828\beta$$

Values of ρ_i (in bold) and P_{ij} are summarized in the following table:

	1	2	3
1	**1.0000**	.7071	
2	.7071	**1.0000**	.7071
3		.7071	**1.0000**

Free valence indices:

$$\mathcal{F}_1 = 1.025;\ \mathcal{F}_2 = 0.318;\ \mathcal{F}_3 = 1.025$$

b. Input Hückel matrix for 90° rotation (transition structure, $H_{2,3} = 0$):

$$\begin{bmatrix} X & 1 & 0 \\ 1 & X & 0 \\ 0 & 0 & X \end{bmatrix} = 0$$

Table of molecular orbitals and energy levels:

	MO 1	MO 2	MO 3
c\E	1.000	.000	-1.000
1	.707	.000	-.707
2	.707	.000	.707
3	.000	1.000	.000

With 2, 1, and 0 electrons in the 3 MOs,

$$E_{\pi} = 3\alpha + 2.000\beta$$

Values of ρ_i (in bold) and P_{ij} are summarized in the following table:

	1	2	3
1	**1.0000**	1.0000	
2	1.0000	**1.0000**	.0000
3		.0000	**1.0000**

Free valence indices:

$$\mathcal{F}_1 = 0.732; \mathcal{F}_2 = 0.732; \mathcal{F}_3 = 1.732$$

Before rotation, E_π for the allyl radical is $3\alpha + 2.828\beta$. After rotation, E_π is $3\alpha + 2.000\beta$. The energy difference is 0.828β. Therefore, the activation energy should be $0.828 \times 18\,\text{kcal/mol} \approx 15\,\text{kcal/mol}$. This result is close to values from more advanced calculations. Li, Z.; Bally, T.; Houk, K. N.; et al. *J. Org. Chem.* **2016**, *81*, 9576 used a quantum mechanical method (G4) developed for the calculation of accurate thermochemical data.[1] For the allyl radical, the rotational barrier was found to be 14 kcal/mol.

For the benzyl system, the delocalization energy of the benzyl radical is 0.721β as obtained by the following HMO calculation and subtracting the π energy from $7\alpha + 8\beta$ ($6\alpha + 8\beta$ for an aromatic ring and α for an electron in a non-interacting p orbital).

Hückel matrix for benzyl radical:

$$\begin{bmatrix} X & 1 & 0 & 0 & 0 & 1 & 1 \\ 1 & X & 1 & 0 & 0 & 0 & 0 \\ 0 & 1 & X & 1 & 0 & 0 & 0 \\ 0 & 0 & 1 & X & 1 & 0 & 0 \\ 0 & 0 & 0 & 1 & X & 1 & 0 \\ 1 & 0 & 0 & 0 & 1 & X & 0 \\ 1 & 0 & 0 & 0 & 0 & 0 & X \end{bmatrix} = 0$$

Table of molecular orbitals and energy levels:

	MO 1	MO 2	MO 3	MO 4	MO 5	MO 6	MO 7
c\E	2.101	1.259	1.000	.000	-1.000	-1.259	-2.101
1	.500	-.500	.000	.000	.000	.500	.500
2	.406	-.116	-.500	-.378	-.500	-.116	-.406
3	.354	.354	-.500	.000	.500	-.354	.354
4	.337	.562	.000	.378	.000	.562	-.337
5	.354	.354	.500	.000	-.500	-.354	.354
6	.406	-.116	.500	-.378	.500	-.116	-.406
7	.238	-.397	.000	.756	.000	-.397	-.238

With 2, 2, 2, 1, 0, 0, and 0 electrons in the 7 MOs,

$$E_\pi = 7\alpha + 8.721\beta$$

Values of ρ_i (in bold) and P_{ij} are summarized in the following table:

	1	2	3	4	5	6	7
1	**1.0000**	.5226				.5226	.6350
2	.5226	**1.0000**	.7050				
3		.7050	**1.0000**	.6350			
4			.6350	**1.0000**	.6350		
5				.6350	**1.0000**	.7050	
6	.5226				.7050	**1.0000**	
7	.6350						**1.0000**

[1] Curtiss, L. A.; Redfern, P. C.; Raghavachari, K. *J. Chem. Phys.* **2007**, *126*, 084108.

Free valence indices:

$\mathcal{F}_1 = 0.052$; $\mathcal{F}_2 = 0.504$; $\mathcal{F}_3 = 0.392$; $\mathcal{F}_4 = 0.462$; $\mathcal{F}_5 = 0.392$;
$\mathcal{F}_6 = 0.504$; $\mathcal{F}_7 = 1.097$

The benzyl radical rotational barrier should be 0.72β or about 13 kcal/mol. Li, Z.; Bally, T.; Houk, K. N.; et al. *J. Org. Chem.* **2016**, *81*, 9576 found the barrier to be 11 kcal/mol for the benzyl radical. Again, the HMO result for the radical was close to the results of a more advanced computational method.

2.16. Using energy levels for benzene and ethene as the energy levels for the transition structure, the DE_π is $(8\alpha + 10.4\beta) - (8\alpha + 10\beta) = 0.4\beta =$ ca. 7 kcal/mol. A literature value is ca. 3 kcal/mol: Sancho-García, J. C.; Pérez-Jiménez, A. J. *J. Phys. B. At. Mol. Opt. Phys.* **2002**, *35*, 1509.

2.17. Binsch, G. *Top. Stereochem.* **1968**, *3*, 97 (see especially pp. 134–146). Shift of electron density from the three-membered ring to the five-membered ring (as shown below) results in a resonance structure with two aromatic rings. Therefore, there is considerable single bond character to the bond connecting the two rings, and the barrier to rotation is lowered in comparison with normal double bonds. The substituents in structure **30** also stabilize the charges in the zwitterionic resonance structure. For the parent hydrocarbon (calicene), the barrier to rotation is ca. 40 kcal/mol: Ghigo, G.; Shahi, A. R. M.; Gagliardi, L.; et al. *J. Org. Chem.* **2007**, *72*, 2823.

2.18. **a.** The data suggest that the $\mathcal{F}_i$ value for the terminal carbon of heptatrienyl should be smaller than that for pentadienyl, while that for octatetraene should be larger than that for hexatriene. The following figure shows the trends in $\mathcal{F}_i$ values for additional members of each series, which appear to approach each other asymptotically. (Filled circles represent radicals, while open circles represent closed-shell polyenes.)

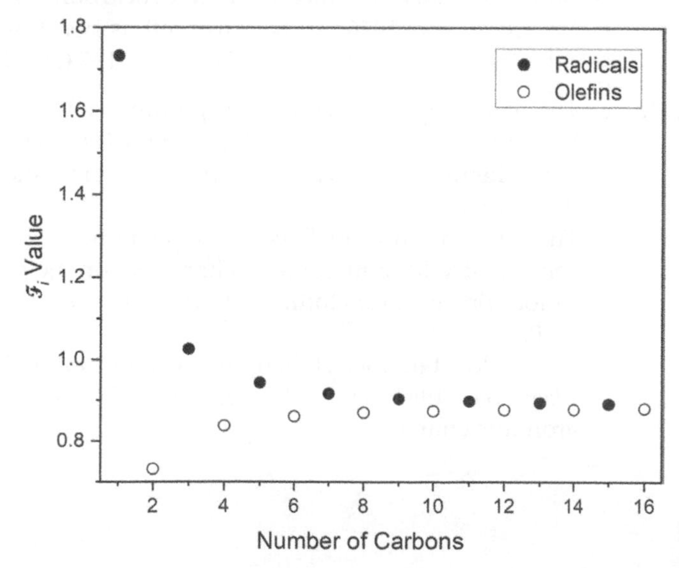

b. Terminal $\mathcal{F}_1$ values are closely related to P_{12} values, which increase as each additional double bond is added to the allyl system. In essence, the terminal $\mathcal{F}_1$ value of a conjugated radical system decreases with increasing chain length because the unpaired electron density is distributed over a larger molecule, meaning that the overlap of the two terminal π centers is more like that in a polyene. In polyenes, terminal $\mathcal{F}_1$ values increase with increasing chain length because P_{12} values decrease (and P_{23} values increase) with increasing chain length.

c. The large value of $\mathcal{F}$ for the methylene carbon atoms of 1,2-dimethylene-3,5-cyclohexadiene can be rationalized with a resonance structure having an aromatic ring and two $-CH_2\bullet$ groups. This model suggests that the two methylene carbon atoms should have considerable radical character, hence large $\mathcal{F}$ values.

d. If COT is tub-shaped, the double bonds should be nearly independent, so the $\mathcal{F}_i$ values should be more similar to those in ethene.

2.19. See the references cited with the problem.

For phenanthrene:

$$RE = 2(5\Gamma_1 + 2\Gamma_2)/5 = 2.32\Gamma_1 = 1.944\,\text{eV}.$$

This compares with 1.933 by SCF-MO and a value of 5.448β or 4.25 eV from HMO (assuming β is 18 kcal/mol = 0.78 eV). See the HMO calculation for phenanthrene above.

For styrene:

$$RE = 2(\Gamma_1 + 0\Gamma_2)/2 = \Gamma_1 = 0.838\,\text{eV}$$

See Herndon, W. C. *J. Am. Chem. Soc.* **1976**, *98*, 887. There are a number of applications for the method, such as calculating the ionization potentials of π-molecular hydrocarbons (Herndon, W. C. *J. Am. Chem. Soc.* **1976**, *98*, 887) as well as bond orders and bond lengths (Herndon, W. C. *J. Am. Chem. Soc.* **1974**, *96*, 7605).

2.20. **a.** The cycloheptatrienylium (tropylium) ion was reported by Doering, W. v. E.; Knox, L. H. *J. Am. Chem. Soc.* **1954**, *76*, 3203. Both chemical reactivity and spectroscopic data suggested that the ion is aromatic.

b. The slow reaction of 5-iodocyclopentadiene with silver ion is consistent with antiaromatic character for the cyclopentadienyl cation. Breslow, R.; Hoffman, J. M., Jr. *J. Am. Chem. Soc.* **1972**, *94*, 2110.

c. The cyclooctatetraenyl dianion was reported by Katz, T. J. *J. Am. Chem. Soc.* **1960**, *82*, 3784, 3785. The NMR spectrum suggested aromatic character.

2.21. The second and fourth drawings represent the same resonance structure. They differ only in the arbitrary placement of the short line beside the longer line of the central double bond.

2.22. a. The three resonance structures for naphthalene are the following:

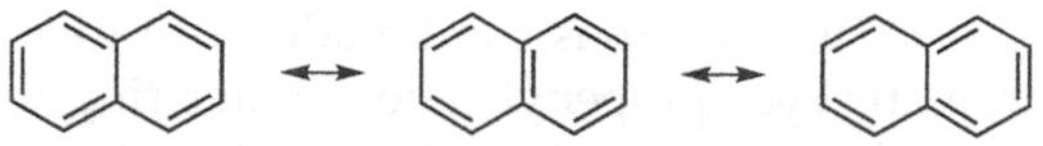

The five resonance structures of phenanthrene are the following:

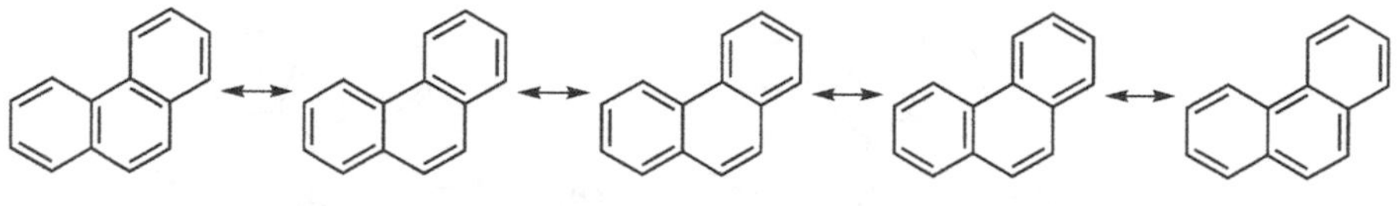

The results can be confirmed by applying the vertex deletion method to each compound.

For naphthalene:

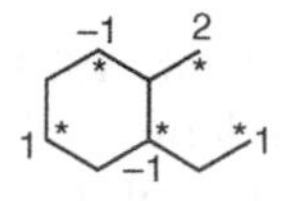

$$2 + 1 = 3$$

For phenanthrene:

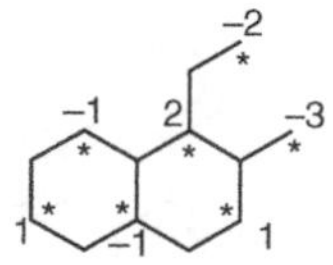

$$|-2| + |-3| = 5$$

b. In phenanthrene, four of the five resonance structures have a double bond between carbon atoms 9 and 10, so the resonance hybrid would be expected to have 80% double bond character between C9 and C10. Electrophilic addition across the 9,10 bond would give a product that retains four of the five resonance structures of the parent compound. In naphthalene, there is a double bond between carbon atoms 1 and 2 in two of the three resonance structures, so it has relatively less olefinic character between these two carbon atoms.

The Clar structure for phenanthrene indicates the double bond character of the 9,10 bond. However, the Clar structure for naphthalene implies aromatic character for the 9,10 positions.

2.23. a. Nine. They are shown on page 51 of Gutman, I.; Cyvin, S. J. *Introduction to the Theory of Benzenoid Hydrocarbons*; Springer-Verlag: Berlin, 1989.

b. Seven. See Gutman, I.; Gojak, S.; Furtula, B.; et al. *Monatsh. Chem.* **2006**, *137*, 1127.

2.24. $(7.435 \times 2) - (0.604 \times 2) - (32.175 \times 2) - (29.38 \times 0) = -50.69$ kcal/mol
The experimental heat of formation of glyoxal is −50.66 kcal/mol (Cox, J. D.; Pilcher, G. *Thermochemistry of Organic and Organometallic Compounds*; Academic Press: New York, 1970; p. 223 and references therein to unpublished work).

2.25. Iyoda, M.; Kurata, H.; Oda, M.; et al. *Angew. Chem. Int. Ed. Engl.* **1993**, *32*, 89. Two electrons can be added to give a dianion that has an aromatic cyclopropenyl cation center ring with each of its carbons bonded to an aromatic cyclopentadienyl anion ring. The stability of the dianion results from the stability of the aromatic rings.

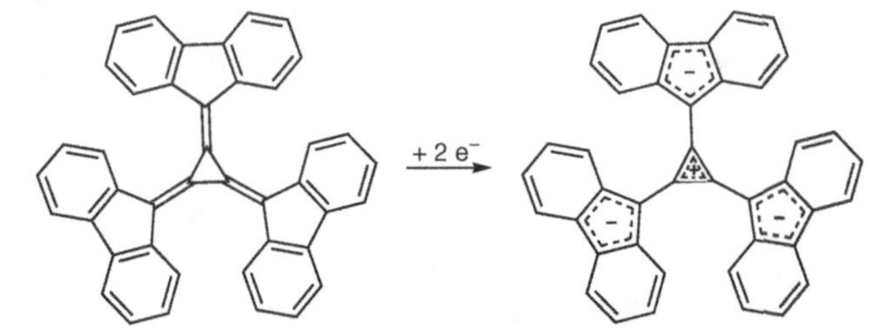

2.26. Baird, N. C. *J. Chem. Educ.* **1971**, *48*, 509. (See especially p. 511.) The DRE for triphenylene is 65 kcal/mol. Three times the DRE for benzene is 63 kcal/mol. By this analysis, the central ring is not aromatic. A similar conclusion is suggested by the Clar structure.

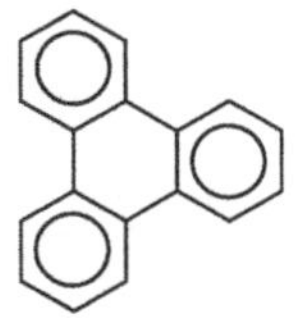

2.27. HOMO for naphthalene is at 0.618β, while that for azulene is 0.400β. Thus, azulene should have a lower ionization potential. For a discussion of such CT complexes, see Frey, J. E.; Andrews, A. M.; Combs, S. D.; et al. *J. Org. Chem.* **1992**, *57*, 6460.

2.28. The HOMO-LUMO gap is smaller in **34** because it is more conjugated. Allinger, N. L.; Siefert, J. H. *J. Am. Chem. Soc.* **1975**, *97*, 752.

2.29. Kitagawa, T.; Takeuchi, K. *J. Phys. Org. Chem.* **1998**, *11*, 157. The reaction is facilitated by formation of the aromatic cyclopropenyl cation and the resonance-stabilized α,α-dicyanobenzyl anion.

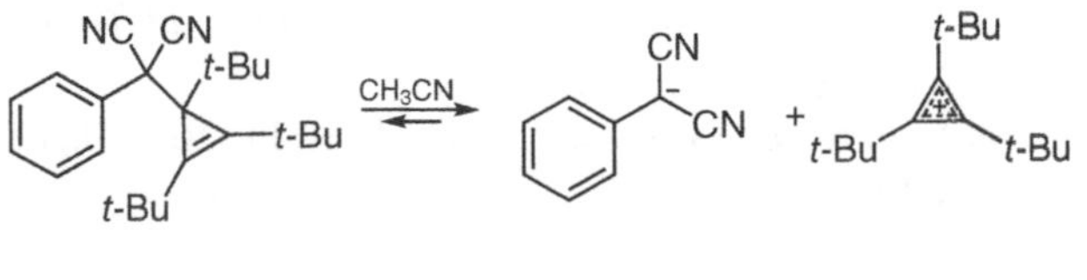

2.30. Structure **37** should be more reactive because its Clar structure has only four sextets and three "double bonds," while the Clar structure for **36** has five sextets and no "double bonds." For a discussion, see page 94 of Gutman, I.; Cyvin, S. J. *Introduction to the Theory of Benzenoid Hydrocarbons*; Springer-Verlag: Berlin, 1989.

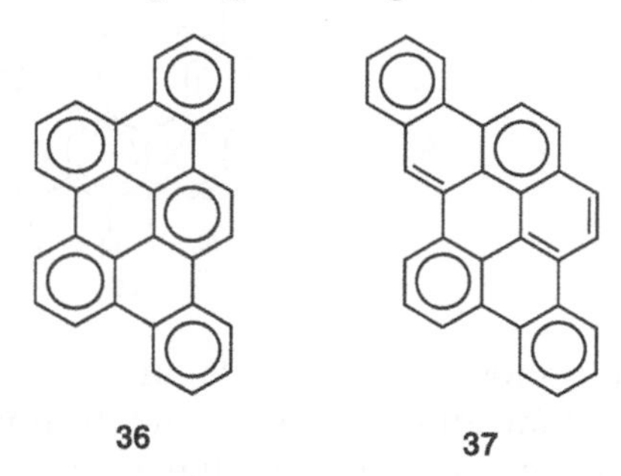

2.31. The larger dipole moment belongs to **39**. Polarization of electrons into the five-membered ring gives it aromatic character like cyclopentadienyl anion. That electron transfer leads to aromatic, cyclopropenyl cation character in the three-membered ring of **39**. In **38**, that electron transfer would produce a nonaromatic four-membered ring, making the polarization less favorable.

2.32. Hirsch, A.; Brettreich, M. *Fullerenes: Chemistry and Reactions*; Wiley-VCH: Weinheim, 2005; pp. 401–402. The criterion that cannot be applied is a pattern of giving electrophilic substitution instead of electrophilic addition, since there are no hydrogen atoms that can be replaced by, e.g., halogens.

2.33. Breslow, R. *Chem. Rec.* **2014**, *14*, 1174. In the first two compounds, the cyclopropyl cyanide should exchange faster because the anion produced by proton removal is not antiaromatic. In contrast, removal of a proton from the cyclopropenyl cyanide produces an antiaromatic ion. For the second pair, the saturated ring should again react faster because iodide loss would produce a nonaromatic cyclopentyl cation. The 5-iodocyclopentadiene is unreactive under the same conditions because iodide loss would produce an antiaromatic cyclopentadienyl cation.

2.34. Dahlstrant, D.; Rosenberg, M.; Kilså, K.; et al. *J. Phys. Chem. A* **2012**, *116*, 5008. The directions of the dipole moments can be predicted on the basis of resonance structures that have aromatic character in one or more of the rings. The relative magnitudes of the dipole moments are a function of the extent of partial charge development and the distance between the partial negative charge and the positive charge.

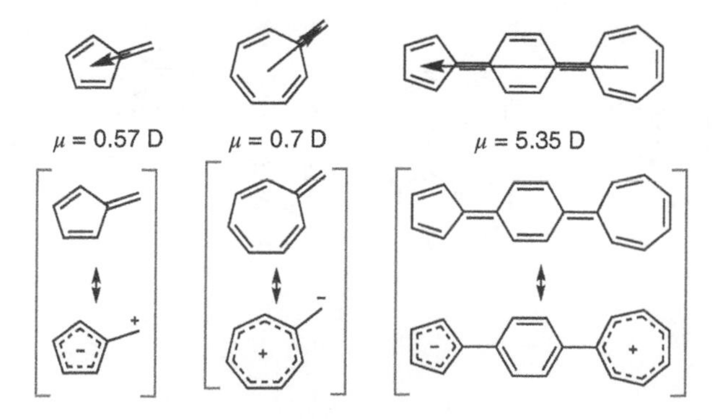

2.35. **a.** There is a maximum of 6 Clar sextets in the structure.

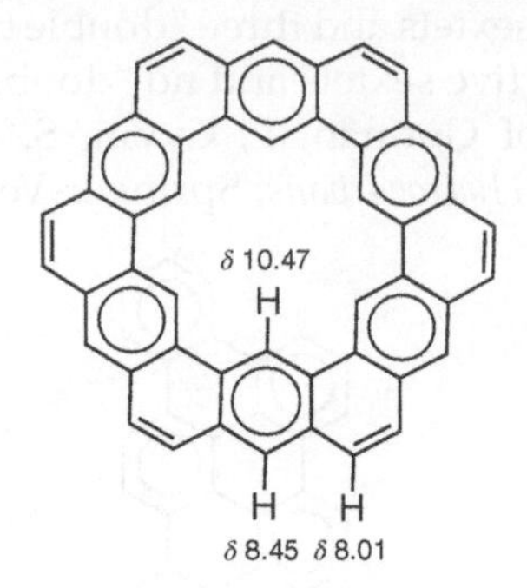

b. The proton chemical shifts are assigned on the basis of anisotropic shielding of protons and the number of hydrogens in each of two environments on the exterior of the ring. Buttrick, J. C.; King, B. T. *Chem. Soc. Rev.* **2017**, *47*, 7. See also Pozo, I.; Majzik, Z.; Pavliček, N.; et al. *J. Am. Chem. Soc.* **2019**, *141*, 15488; Haags, A.; Reichmann, A.; Fan, Q.; et al. *ACS Nano* **2020**, *14*, 15766.

2.36. Anjalikrishna, P. K.; Suresh, C. H.; Gadre, S. R. *J. Phys. Chem. A* **2019**, *123*, 10139. Naphthalene (1), anthracene (1), chrysene (2), perylene (2), tetracene (1), benzanthrene (2), coronene (3).

2.37. Anjalikrishna, P. K.; Suresh, C. H.; Gadre, S. R. *J. Phys. Chem. A* **2019**, *123*, 10139. No. Each of the two resonance structures here has three Clar circles, but neither has a Clar circle in the center ring. It is considered an "empty ring."

2.38. Wiberg, K. B. *Chem. Rev.* **2001**, *101*, 1317. Cyclopropane is more acidic than cyclopropene because the cyclopropenyl anion is antiaromatic; cyclopentadiene is more acidic than 1,4-pentadiene because the cyclopentadienyl anion is aromatic; cyclononatrienyl anion is more acidic than cycloheptatrienyl anion because cyclononatrienyl anion is aromatic.

2.39. Koper, C.; Sarobe, M.; Jenneskens, L. W. *Phys. Chem. Chem. Phys.* **2004**, *6*, 319. Structures (a) and (b) are alternant; (c) and (d) are nonalternant.

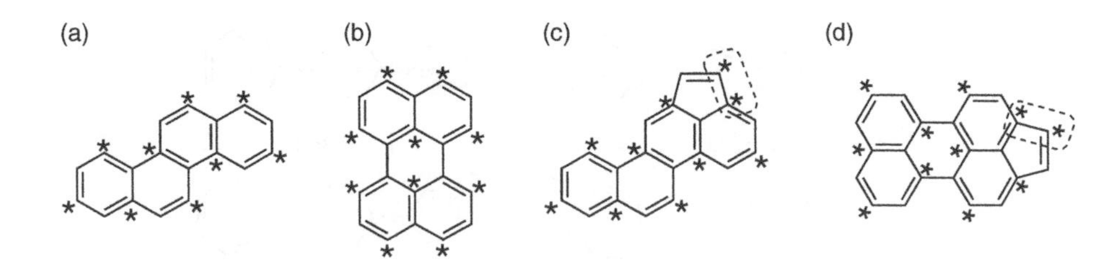

2.40. Ghosh, D.; Periyasamy, G.; Pati, S. K. *Phys. Chem. Chem. Phys.* **2011**, *13*, 20627. The Clar structure makes clear that three of the rings have strong aromatic character but the central ring does not. Therefore, the three outer rings should have the more negative NICS(0) value (−7.83) and the larger HOMA value (0.92).

2.41. Pavliček, N.; Mistry, A.; Majzik, Z.; et al. *Nature Nanotech.* **2017**, *12*, 308. [3]Triangulene has diradical character because it is not possible to draw a structure with only double or single bonds and no unpaired electrons.

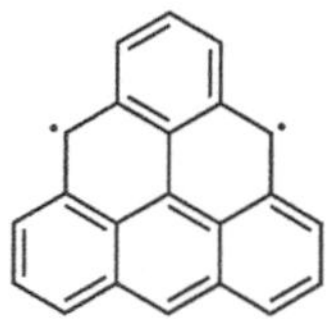

2.42. Most organic chemists would have difficulty abandoning the idea of a bond because Lewis theory represented with line formulas is the historical framework on which structure theory was built. Numbers alone do not lend themselves to an easy conceptual understanding of transferrable chemical concepts.

2.43. The statement in the question was paraphrased from Shaik, S. *New J. Chem.* **2007**, 81, 2015. A numerical answer from a calculation should always be tested for reasonableness in terms of accepted chemical models. Otherwise, the calculation does not provide the chemist with a better understanding of those models.

2.44. A theoretical result that agrees with a well-known experimental result has little value unless the theory behind the calculation can be used to predict the result of a new experiment. The computational chemist must establish theoretical methods with carefully validated data sets and then predict results for new experiments.

2.45. Benzene and aromaticity are inseparable because the properties of benzene set the standard for determining what is aromatic and, in turn, benzene necessarily matches those properties better than does any other compound.

2.46. Even though MO theory assumes that orbitals are delocalized over an entire structure and that any part of a structure is affected by the rest of it, localized orbital approximations common to different functional groups have utility in contemporary chemistry. In that sense, the idea of a functional group is consistent with molecular orbital theory.

2.47. The statement that bond-length-alternating 1,3,5-cycylohexatriene would be nearly as aromatic as benzene assumes that aromaticity is purely a question of relative thermodynamic stabilities (e.g. of cyclic and acyclic structures). However, having equal carbon–carbon bond lengths was one of the early criteria for aromatic character. The equal carbon–carbon bond lengths of benzene are unlikely to be found in any structure other than benzene itself, but that does not mean that benzene is the exception to what it means to be aromatic. It would be just as reasonable to say that benzene is the only structure to be fully aromatic.

2.48. The *concept* of aromaticity certainly exists, even if it is only a concept, because it helps rationalize important patterns of chemical structure and reactivity.

CHAPTER 3

Stereochemistry

3.1. **a.** (*Z*)
b. *trans*; (1*R*,2*R*)
c. (1*Z*,3*Z*)
d. (*Z*) or *syn*
e. *exo*
f. *rac-trans*
g. (1*R**,3*R**,5*R**)
h. *ß*

3.2. **a.** *E*.
b. *Z*.

3.3. Structures (a) and (b) are positional isomers because the methoxy group is on C2 in (a) but is on C1 in (b). Structures (a) and (c) are enantiomers. Structures (a) and (d) are functional group isomers because (a) is a methyl ether, while (d) is an alcohol. Structures (a) and (e) are diastereomers.

3.4. **a.** (*R*)-2-chloro-3-oxopropanoic acid
b. (*S*)-3-bromo-3-phenylpropene
c. (3*R*,4*S*)-3-bromo-4-chloro-3-methylhexane
d. (2*S*,3*S*)-2-methoxy-2,3-butanediol
e. (2*R*,3*S*)-4-chloro-2,3-dihydroxybutanal
f. (*S*)-1,3-dichloro-1,2-propadiene
g. (1*S*,2*S*)-1,2-dichlorocyclohexane
h. (*R*)-2-bromo-2′-chlorobiphenyl
i. (1*R*,2*R*)-1,2-dimethylcyclopropane

Solutions Manual for Perspectives on Structure and Mechanism in Organic Chemistry, Third Edition.
Felix A. Carroll.

j. (*R*)-1,2-cyclononadiene
Moore, W. R.; Anderson, H. W.; Clark, S. D.; et al. *J. Am. Chem. Soc.* **1971**, *93*, 4932.

k. (*R*)-1-(Bromomethylidene)-4-methylcyclohexane
Gerlach, H. *Helv. Chim. Acta* **1966**, *48*, 1291.

l. (*R*)-6,6′-dimethyl-2,2′-diphenic acid
Newman, P.; Rutkin, P.; Mislow, K. *J. Am. Chem. Soc.* **1958**, *80*, 465.

m. (2*S*,4*R*,5*S*)-2,5-diamino-3,3-difluoro-1,6-diphenylhexan-4-ol
Sham, H. L.; Wideburg, N. E.; Spanton, S. G.; et al. *J. Chem. Soc. Chem. Commun.* **1991**, 110. The structure has a different name in the original publication.

3.5. a. (1*R*,2*R*,3*R*,4*S*) [The NH_2 group is a substituent on C2.]
Chamberlain, B. T.; Vincent, M.; Nafie, J.; et al. *J. Org. Chem.* **2021**, *86*, 4281.

b. (2*Z*,4*E*,6*R*,7*S*,9*S*,12*S*,13*R*,14*S*,16*S*,19*R*,20*S*,21*S*,22*S*,23*Z*)
Paterson, I.; Britton, R.; Delgado, O.; et al. *Chem. Commun.* **2004**, 632.

c. (1*R*,6*R*,7*S*,10*R*,11*R*,14*S*,15*R*)
Chittiboyina, A. G.; Kumar, G. M.; Carvalho, P. B.; et al. *J. Med. Chem.* **2007**, *50*, 6299.

d. (4*R*,5*R*,6*E*,9*S*,11*Z*,14*R*,15*E*,18*S*,19*R*,21*S*,22*S*)
Tsuda, M.; Oguchi, K.; Iwamoto, R.; et al. *J. Nat. Prod.* **2007**, *70*, 1661.

e. (3*S*,2″*S*,5″*S*,7″*S*)
Brimble, M. A.; Bryant, C. J. *Org. Biomol. Chem.* **2007**, *5*, 2858.

f. (3*E*,5*S*,8*R*,11*E*,13*S*,16*R*)
Yadav, J. S.; Subba Reddy, U. V.; Subba Reddy, B. V. *Tetrahedron Lett.* **2009**, 5984.

g. (3*R*,4*S*,5*S*)
Burks, E. A.; Johnson, W. H., Jr.; Whitman, C. P. *J. Am. Chem. Soc.* **1998**, *120*, 7665.

h. (*R*)
McGrath, M. J.; Fletcher, M. T.; König, W. A.; et al. *J. Org. Chem.* **2003**, *68*, 3739.

i. (1*R*,4*R*,6*R*,9*R*)
McCann, D. M.; Stephens, P. J. *J. Org. Chem.* **2006**, *71*, 6074.

j. (1*R*,5*S*,8*S*,9*S*,10*S*)
Stephens, P. J.; Pan, J. J.; Devlin, F. J.; et al. *J. Org. Chem.* **2007**, *72*, 3521.

k. (1*R*,4*S*,8*R*,11*S*)
McCann, D. M.; Stephens, P. J. *J. Org. Chem.* **2006**, *71*, 6074.

l. (*R*)
Senanayake, C. H.; Krishnamurthy, D.; Lu, Z.-H.; et al. *Aldrichchimica Acta* **2005**, *38*, 93.

m. (2*R*,3*R*,4*R*)
Mennucci, B.; Claps, M.; Evidente, A.; et al. *J. Org. Chem.* **2007**, *72*, 6680.

n. (1*R*,2*S*,7*S*,8*R*)
Kuppens, T.; Vandyck, K.; Van der Eycken, J.; et al. *J. Org. Chem.* **2005**, *70*, 9103.

o. (*R*)
Chan, J. Y. C.; Hough, L.; Richardson, A. C. *J. Chem. Soc. Perkin Trans. 1* **1985**, 1457.

p. (2*R*,3*R*,4*R*,5*R*)
Reddy, J. S.; Rao, B. V. *J. Org. Chem.* **2007**, *72*, 2224.

q. (5*R*,10*R*,16*R*,18*S*,19*R*,20*S*)
Bonazzi, S.; Güttinger, S.; Zemp, I.; et al. *Angew. Chem. Int. Ed.* **2007**, *46*, 8707.

r. (1*S*,2*R*)
Rudchenko, V. F.; Dyachenko, O. A.; Zolotoi, A. B.; et al. *Tetrahedron* **1982**, *38*, 961.

s. (*R*)

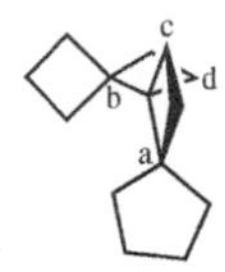

Widjaja, T.; Fitjer, L.; Pal, A.; et al. *J. Org. Chem.* **2007**, *72*, 9264.

t. (*S*)

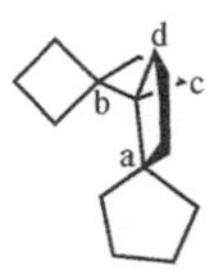

Widjaja, T.; Fitjer, L.; Pal, A.; et al. *J. Org. Chem.* **2007**, *72*, 9264.

3.6. a. (*P*)
Bringmann, G.; Heubes, M.; Breuning, M.; et al. *J. Org. Chem.* **2000**, *65*, 722.

b. (*M*)
Goel, A.; Singh, F. V.; Kumar, V.; et al. *J. Org. Chem.* **2007**, *72*, 7765.

3.7. Because the helix makes a clockwise turn as it proceeds away from the observer, it is a *P* helix, which corresponds to the *S* designation.

3.8. a. *erythro*
Yang, H.; Liebeskind, L. S. *Org. Lett.* **2007**, *9*, 2993.

b. *threo*
Hayashi, S.; Hirano, K.; Yorimitsu, H.; et al. *Org. Lett.* **2005**, *7*, 3577.

3.9. 4096. The number of possible stereoisomers is 2^n, where n is the number of chiral centers. There are twelve chiral centers, each marked with a star in the figure. Seebach, D.; Lapierre, J.-M.; Skobridis K.; et al. *Angew. Chem. Int. Ed. Engl.* **1994**, *33*, 440.

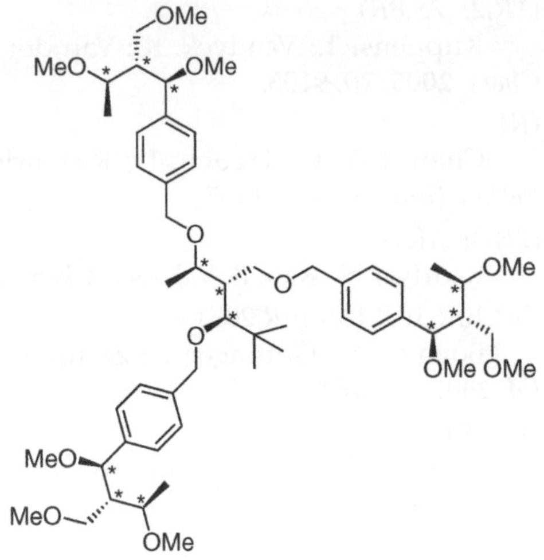

3.10.

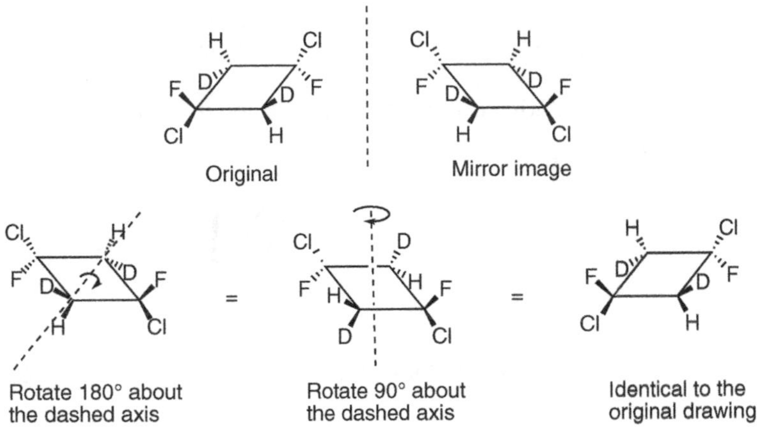

Original

Mirror image

Rotate 180° about the dashed axis

Rotate 90° about the dashed axis

Identical to the original drawing

3.11. 16.

Dandapani, S.; Jeske, M.; Curran, D. P. *J. Org. Chem.* **2005**, *70*, 9447.

3.12. **a.** Cywin, C. L.; Webster, F. X.; Kallmerten, J. *J. Org. Chem.* **1991**, *56*, 2953.

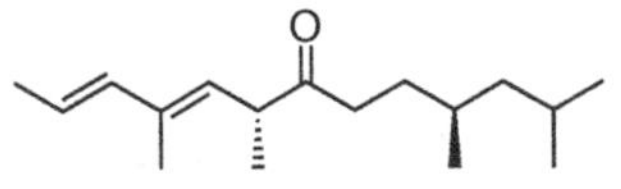

b. Ingold, K. U. *Aldrichimica Acta* **1989**, 22, 69.

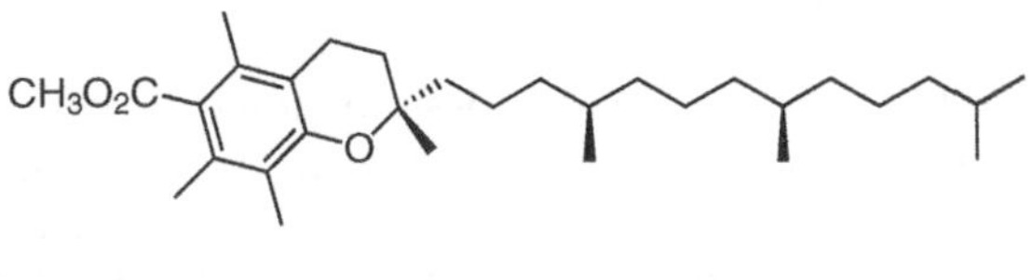

c. Rychnovsky, S. D.; Griesgraber, G.; Zeller, S.; et al. *J. Org. Chem.* **1991**, *56*, 5161.

d. Cianciosi, S. J.; Ragunathan, N.; Freedman, T. B.; et al. *J. Am. Chem. Soc.* **1990**, *112*, 8204.

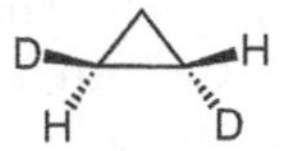

e. This structure was illustrated in an Eastman Fine Chemical advertisement in *J. Org. Chem.* **1992**, *57* (20) on a page preceding the table of contents.

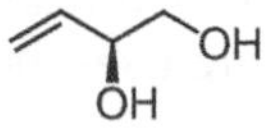

f. This structure was illustrated in an Eastman Fine Chemicals advertisement in *J. Org. Chem.* **1992**, *57* (20) on a page preceding the table of contents.

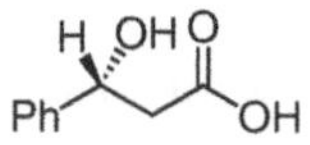

g. Bharucha, K. N.; Marsh, R. M.; Minto, R. E.; et al. *J. Am. Chem. Soc.* **1992**, *114*, 3120.

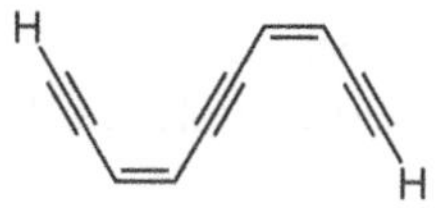

h. Naoshima, Y.; Munakata, Y.; Yoshida, S.; et al. *J. Chem. Soc. Perkin Trans. 1* **1991**, 549.

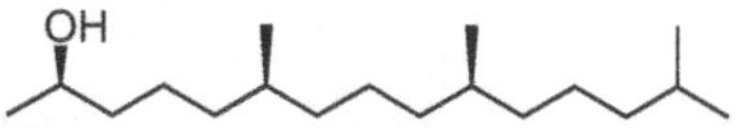

i. Walborsky, H. M.; Goedken, V. L.; Gawronski, J. K. *J. Org. Chem.* **1992**, *57*, 410.

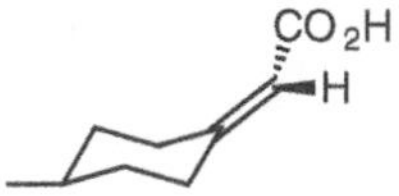

j. Rawson, D.; Meyers, A. I. *J. Chem. Soc. Chem. Commun.* **1992**, 494.

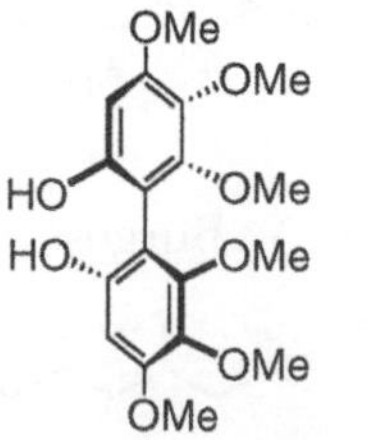

k. Freedman, T. B.; Cianciosi, S. J.; Ragunathan, N.; et al. *J. Am. Chem. Soc.* **1991**, *113*, 8298.

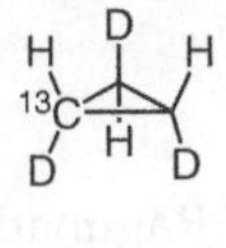

l. Liu, C.; Coward, J. K. *J. Org. Chem.* **1991**, *56*, 2262. The name 2-(phenylmethoxy)ethyl means that there is a $C_6H_5CH_2OCH_2CH_2$ substituent on C2. An alternative name is (*S*)-2-(benzyloxy) ethyl-2-methyloxirane.

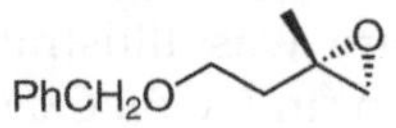

m. Glattfeld, J. W. E.; Chittum, J. W. *J. Am. Chem. Soc.* **1933**, *55*, 3663. The drawing here uses the Maehr notation.

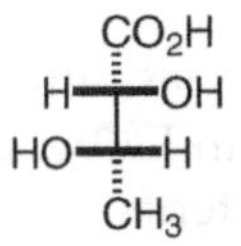

n. Moorthy, J. N.; Venkatesan, K. *J. Org. Chem.* **1991**, *56*, 6957.

3.13. a. Cavagnat, D.; Lespade, L.; Buffeteau, T. *J. Phys. Chem. A.* **2007**, *111*, 7014.

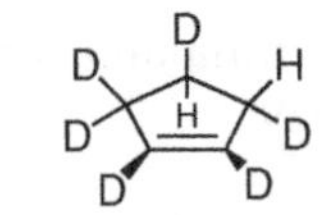

b. Griesbeck, A. G.; Miara, C.; Neudörfl, J. *ARKIVOC* **2007**, *viii*, 216.

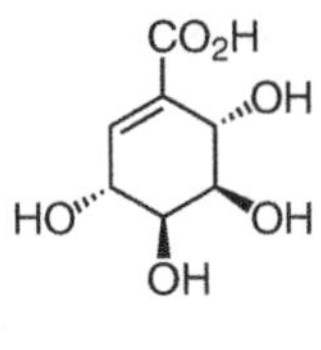

c. Gao, X.; Lin, C.-J.; Jia, Z.-J. *J. Nat. Prod.* **2007**, *70*, 830.

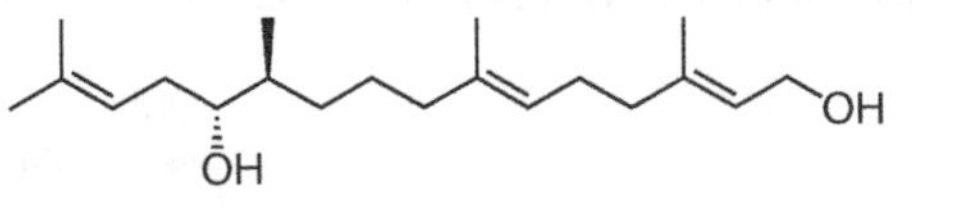

d. Zhou, J.; Zhu, Y.; Burgess, K. *Org. Lett.* **2007**, *9*, 1391.

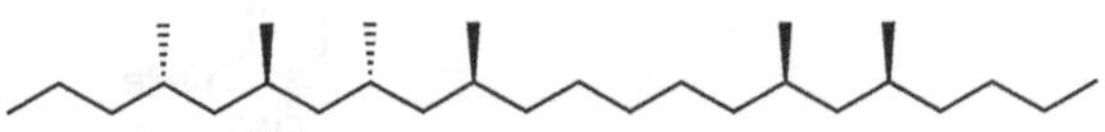

e. Kang, B.; Britton, R. *Org. Lett.* **2007**, *9*, 5083.

O

f. Otomaru, Y.; Tokunaga, N.; Shintani, R.; et al. *Org. Lett.* **2005**, *7*, 307.

g. Vidal-Cros, A.; Gaudry, M.; Marquet, A. *J. Org. Chem.* **1985**, *50*, 3163.

CO_2H
H_2N—H
H—F
CH_2CO_2H

h. Burton, G. W.; de la Mare, P. B. D.; Wade, M. *J. Chem. Soc. Perkin Trans. 2* **1974**, 591.

Cl
Cl
Cl
Cl

i. Rudchenko, V.; Dyachenko, O. A.; Zolotoi, A. B.; et al. *Tetrahedron* **1982**, *38*, 961.

CH_3
N
Cl
CH_3

j. Nawrath, T.; Dickshat, J. S.; Müller, R.; et al. *J. Am. Chem. Soc.* **2008**, *130*, 430.

H

k. Cho, E. J.; Lee, D. *Org. Lett.* **2008**, *10*, 257.

OH
OH
OH

l. Andersen, K. K.; Colonna, S.; Stirling, C. J. M. *J. Chem. Soc. Chem. Commun.* **1973**, 645.

O
$Ph^{12}CH_2$ $^{13}CH_2Ph$

m. Kitching, W.; Lewis, J. A.; Perkins, M. V.; et al. *J. Org. Chem.* **1989,** *54*, 3893.

n. Rao, A. V. R.; Gurjar, M. K.; Bose, D. S.; et al. *J. Org. Chem.* **1991,** *56*, 1320.

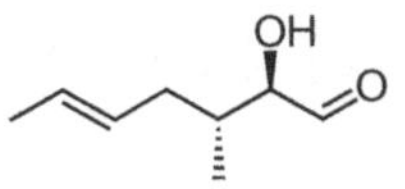

OH O

o. Chattopadhyay, S.; Mamdapur, V. R.; Chadha, M.S. *J. Chem. Res. (M)* **1990,** 1818.

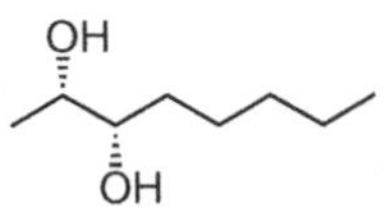

OH OH

p. King, S. B.; Ganem, B. *J. Am. Chem. Soc.* **1994,** *116*, 562.

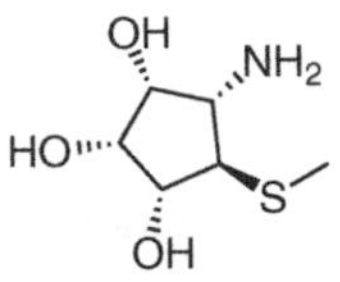

OH NH_2 HO S OH

3.14. These problems require making a 3-D drawing of each reactant and each product before analyzing for inversion or retention. Decisions cannot be made only on the basis of (*R*), (*S*), (+), or (−) designations.

a. Coke, J. L.; Shue, R. S. *J. Org. Chem.* **1973,** *38*, 2210. Retention.

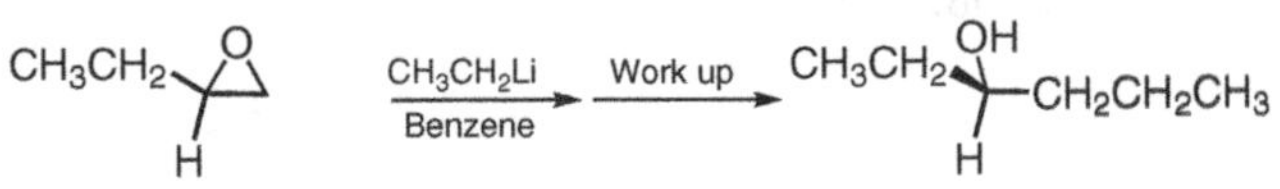

CH_3CH_2 O H → CH_3CH_2Li / Benzene → Work up → CH_3CH_2 OH $CH_2CH_2CH_3$ H

(*R*)-(+)-1,2-expoxybutane (*R*)-(−)-hexan-3-ol

b. Katsura, T.; Minamii, M. 61,176,557; see *Chem. Abstr.* **1986,** *106*, 66799f; for a discussion, see Salaün, J. *Chem. Rev.* **1989,** *89*, 1247. Retention.

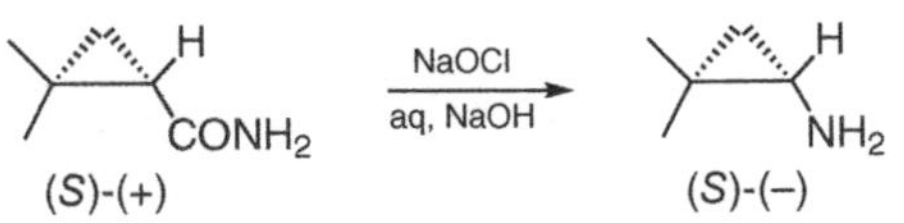

H $CONH_2$ → NaOCl / aq, NaOH → H NH_2

(*S*)-(+) (*S*)-(−)

c. Floss, H. G.; Lee, S. *Acc. Chem. Res.* **1993,** *26*, 116. Inversion.

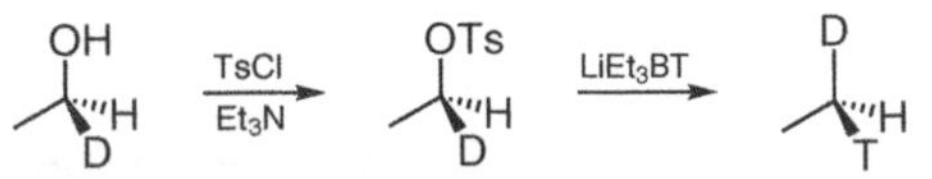

OH H D → TsCl / Et_3N → OTs H D → $LiEt_3BT$ → D H T

(*S*) (*S*) (*S*)

d. Floss, H. G.; Lee, S. *Acc. Chem. Res.* **1993**, *26*, 116. Retention,

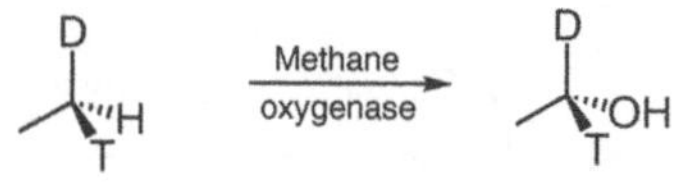

Methane oxygenase

(*S*) (*R*)

e. Walborsky, H. M.; Impastato, F. J.; Young, A. E. *J. Am. Chem. Soc.* **1964**, *86*, 3283. Retention.

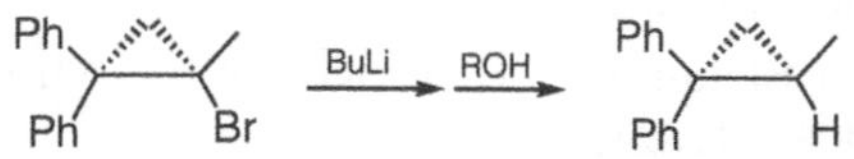

BuLi ROH

(*S*)-(+) (*R*)-(−)

f. Skell, P. S.; Pavlis, R. R.; Lewis, D. C.; et al. *J. Am. Chem. Soc.* **1973**, *95*, 6735. Inversion at C2 and retention at C3.

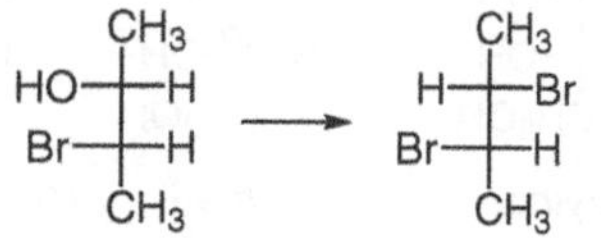

(2*R*,3*S*) (2*S*,3*S*)

g. Wiberg, K. B. *J. Am. Chem. Soc.* **1952**, *74*, 3891. Retention.

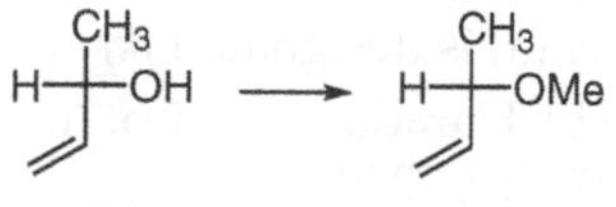

D-(+) D-(−)

h. Abate, A.; Brenna, E.; Fuganti, C.; et al. *J. Org. Chem.* **2005**, *70*, 1281. Retention at the remaining chiral centers.

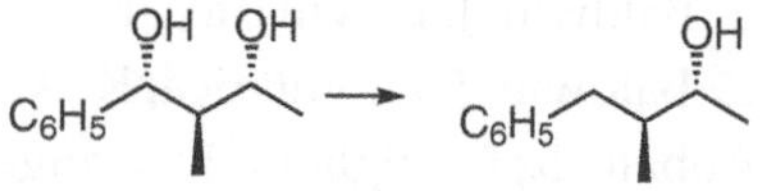

i. McCoull, W.; Davis, F. A. *Synthesis* **2000**, 1347. This product is formed by inversion. The (2*S*,3*S*) diastereomer, formed by retention of configuration, is also a product of the reaction.

j. Jones, W. M.; Wilson, J. W., Jr. *Tetrahedron Lett.* **1965**, 1587. Retention at both chiral centers.

(2*R*,3*R*)-2,3-diphenylcyclopropane-1-carboxylic acid (1*S*,2*S*)-1,2-diphenylcyclopropane

k. Kuwahara, S.; Obata, K.; Fujita, T.; et al. *Eur. J. Org. Chem.* **2010**, 6385. Retention.

CBr_4, PPh_3 / CH_2Cl_2; $NaBH_4$ / HMPA

(*R*)-3-butyl-3-methylhexane-1,6-diol → (*S*)-4-ethyl-4-methyloctane

3.15. a. (R_a) Blakemore, P. R.; Kilner, C.; Milicevic, S. D. *J. Org. Chem.* **2006**, *71*, 8212.

b. (S_a) Blakemore, P. R.; Kilner, C.; Milicevic, S. D. *J. Org. Chem.* **2006**, *71*, 8212.

3.16. L. Hanessian, S.; Auzzas, L. *Org. Lett.* **2008**, *10*, 261.

3.17.

D-*erytho*	D-*threo*	L-*erytho*	L-*threo*
CH_2OH	CH_2OH	CH_2OH	CH_2OH
CH_3–(*S*)–OH	HO–(*R*)–CH_3	HO–(*R*)–CH_3	CH_3–(*S*)–OH
H–(*R*)–OH	H–(*R*)–OH	HO–(*S*)–H	HO–(*S*)–H
CH_2OH	CH_2OH	CH_2OH	CH_2OH

3.18. a. The structure is (S,S_a,S). Blakemore, P. R.; Kilner, C.; Milicevic, S. D. *J. Org. Chem.* **2006**, *71*, 8212.

b. The structure is (S,R_a,S). Therefore, (a) and (b) are diastereomers. Structure (a) is designated *lk*, and structure (b) is designated *ul*.

c. (*M*,*M*)-(*E*). Harada, N.; Saito, A.; Koumura, N.; et al. *J. Am. Chem. Soc.* **1997**, *119*, 7249.

d. The structure is (*P*,*S*). Stará, I. G.; Alexandrová, Z.; Teplý, F.; et al. *Org. Lett.* **2005**, *7*, 2547.

e. (*R*). Cipiciani, A.; Fringuelli, F.; Mancini, V.; et al. *J. Org. Chem.* **1997**, *62*, 3744.

f. *syn-E*. Baldwin, J. E.; Villarica, K. A. *J. Org. Chem.* **1995**, *60*, 186.

g. *syn-Z*. Baldwin, J. E.; Villarica, K. A. *J. Org. Chem.* **1995**, *60*, 186.

h. (*S*). Abbate, S.; Castiglioni, E.; Gangemi, F.; et al. *J. Phys. Chem. A* **2007**, *111*, 7031.

3.19. a. D_3, chiral. Eaton, P. E.; Leipzig, B. *J. Org. Chem.* **1978**, *43*, 2483.

b. C_2, chiral. Halterman, R. L.; Jan, S.-T. *J. Org. Chem.* **1991**, *56*, 5253.

c. C_2, chiral. See the discussion by Deprés, J.-P.; Morat, C. *J. Chem. Educ.* **1992**, *69*, A232.

d. C_2, chiral. Wang, Y.; Stretton, A. D.; McConnell, M. C.; et al. *J. Am. Chem. Soc.* **2007**, *129*, 13193.

e. D_3, chiral. Wang, Y.; Stretton, A. D.; McConnell, M. C.; et al. *J. Am. Chem. Soc.* **2007**, *129*, 13193.

3.20. Hoye, T. R.; Hanson, P. R.; Kovelesky, A. C.; et al. *J. Am. Chem. Soc.* **1991**, *113*, 9369.

a.

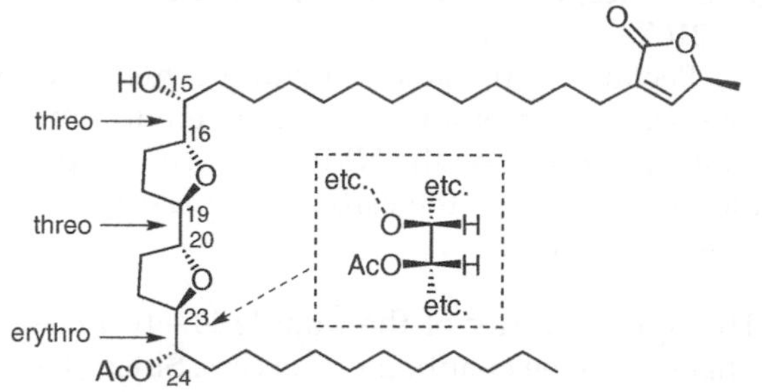

In order to determine the threo and erythro designations, it is helpful to redraw selected portions of the molecule as Fischer projections. One example is shown above.

b.

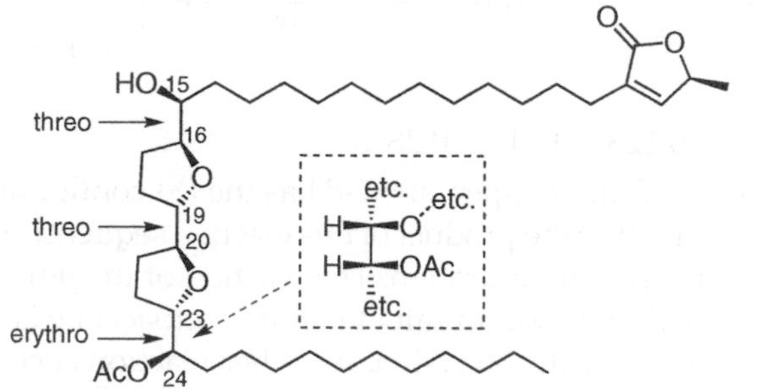

Note that the stereochemical designations do not change because the hexaepi structure is the mirror image of the original structure along the C15–C24 segment. The hexaepi structure is not the enantiomer of the original structure, however, because the configuration of the chiral center in the lactone group is not changed.

3.21. Polniaszek, R. P.; Dillard, L. W. *Abstracts of the 203rd National Meeting of the American Chemical Society*, San Francisco, CA, April 5–10, **1992**, Abstract ORGN 494.

Invert the configuration at C2 (the carbon atom adjacent and to the right of the N) by moving the methyl group to the rear and the propyl group to the front.

3.22. Reynolds, K. A.; Fox, K. M.; Yuan, Z.; et al. *J. Am. Chem. Soc.* **1991**, *113*, 4339. As drawn on the page, the left face of (a) is the *si* face. Similarly, the left face of (b) is the *re* face, and the front face of (c) is the *si* face.

3.23. Reynolds, K. A.; Fox, K. M.; Yuan, Z.; et al. *J. Am. Chem. Soc.* **1991**, *113*, 4339.

The protons in (a), (c), and (d) are heterotopic, stereoheterotopic and enantiotopic, while those in (b) are heterotopic, stereoheterotopic, and diastereotopic. As the structures are drawn, in (a) the circled hydrogen on the left is pro-(*S*). In (b) the circled hydrogen on the left is pro-(*R*). In (c) the circled hydrogen on top is pro-(*S*). In (d) the circled hydrogen in front is pro-(*S*).

3.24. Whitesides, G. M.; Kaplan, F.; Roberts, J. D. *J. Am. Chem. Soc.* **1963**, *85*, 2167.

Because of the adjacent chiral center, the two methylene protons are diastereotopic unless their environments are exchanged at a rate that is fast on the NMR time scale. A process such as reversible dissociation of the carbon–magnesium bond could lead to the observed results.

3.25. The optical purity of the sample is −16.19/−23.13 = 0.70 or 70%. That is also the enantiomeric excess. Since $(1-x)-x=0.70$, $x=0.15$. Therefore, the mole fraction of (−) enantiomer is 85%, while that of (+) enantiomer is 15%.

3.26. Streitwieser, A., Jr.; Granger, M. R. *J. Org. Chem.* **1967**, *32*, 1528.

(*S*)-(−) $H-C(D)(CH_3)-OH$ —[$O_2N-C_6H_4-SO_2Cl$]→ $H-C(D)(CH_3)-OSO_2C_6H_4NO_2$ —[$NaCH_3COCHCO_2CH_3$, CH_3OH]→ —[(1) NaOH, H_2O; (2) Δ ($-CO_2$); (3) Reduce]→ $CH_3CH_2CH_3-C(D)(CH_3)-H$ (*R*)-(+)

a. −0.123°/0.44 = −0.28°;

b. (*S*). Since (−)-pentane-2-*d* has the (*S*) configuration, it must be the case that the product of this reaction sequence, (+)-pentane-2-*d*, has the (*R*) configuration. The reaction of the *p*-nitrobenzenesulfonate with the sodium salt of methyl acetoacetate is an S_N2 reaction that occurs with inversion. All other reactions occur with retention of configuration at the chiral center.

3.27. Hilvert, D.; Nared, K. D. *J. Am. Chem. Soc.* **1988**, *110*, 5593.

a.

CO_2^- / CO_2^- / OH

b. The CD spectra are mirror images.

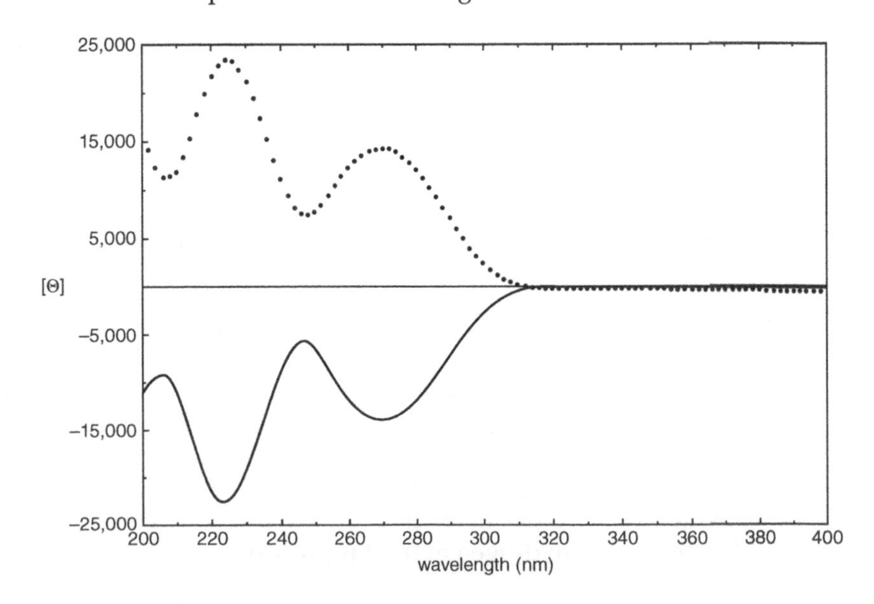

3.28. Structures (a) and (b) are identical, so they would have the same CD spectrum. Structure (c) is the enantiomer of structure (a), so its CD spectrum would be the mirror image of that for structure (a). Xu, Z.-X.; Zhang, C.; Zheng, Q.-Y.; et al. *Org. Lett.* **2007,** *9,* 4447.

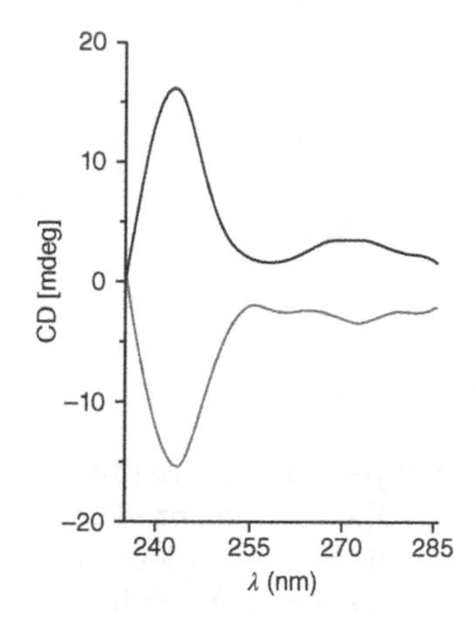

3.29. LeGoff, E.; Ulrich, S. E.; Denney, D. B. *J. Am. Chem. Soc.* **1958,** *80,* 622.

CH_3 / Br–C–H / n-C_6H_{13} (*R*)-(−) $\xrightarrow{a}$ CH_3 / H–C–$CH(CO_2Et)_2$ / n-C_6H_{13} $\xrightarrow{b}$ CH_3 / H–C–CH_2CO_2H / n-C_6H_{13} $\xrightarrow{c}$ CH_3 / H–C–CH_2CH_3 / n-C_6H_{13} (*R*)-(−)

a. $NaCH(CO_2Et)_2$, CH_3CH_2OH
b. (1) KOH, H_2O (2) Δ
c. (1) $LiAlH_4$ (2) PBr_3 (3) $LiAlH_4$

Reaction a is an S_N2 reaction, which proceeds with inversion of configuration. The fact that both the reactant and product of step a are (−) does not indicate retention. Step b is a hydrolysis and decarboxylation. Step c is a reduction of the carboxylic acid to an alcohol, conversion of the alcohol to a bromide, and reduction of the carbon–bromine bond to a carbon–hydrogen bond. Neither step b nor step c involves bond breaking at the chiral center, so both proceed with retention of configuration. Overall, therefore, the reaction sequence converts (*R*)-(−)-2-bromooctane to (*R*)-(−)-3-methylnonane with net inversion of configuration.

3.30. a. In the absence of isotopic labels, the parent reaction is only stereoselective, not stereospecific, because stereoisomeric reactants are not possible.

b. Miyamoto, K.; Tsuchiya, S.; Ohta, H. *J. Am. Chem. Soc.* **1992,** *114,* 6256. The reaction takes place with inversion of configuration and with removal of the pro-(*R*) carboxyl group.

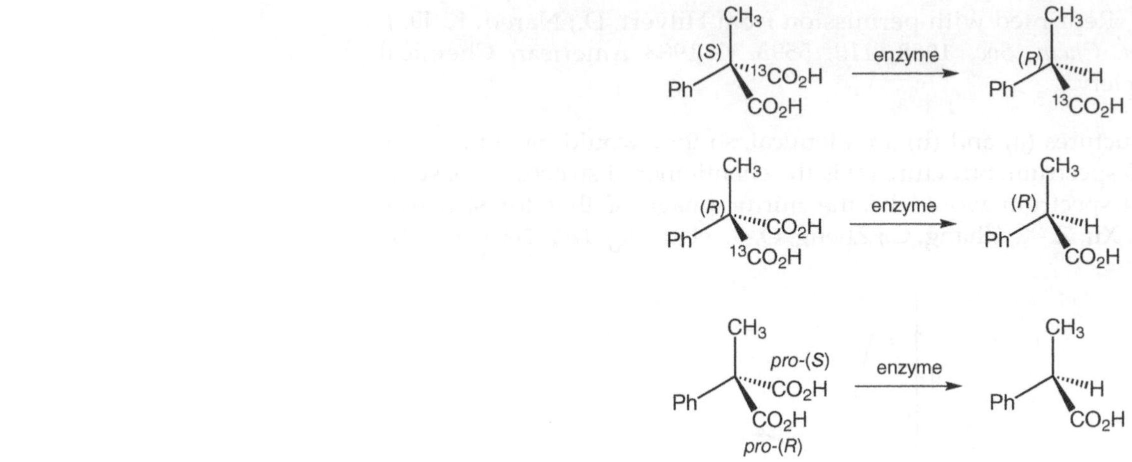

3.31. Cinquini, M.; Cozzi, F.; Sannicolò, F.; et al. *J. Am. Chem. Soc.* **1988**, *110*, 4363.

The second Baeyer–Villager oxidation converts the intermediate structure into either of two products. One of them is reduced with $LiAlH_4$ to two molecules of 2-methyl-1,4-butanediol. Reduction of the other product produces two achiral diols. See the reference for a more detailed discussion of the stereochemistry of these compounds.

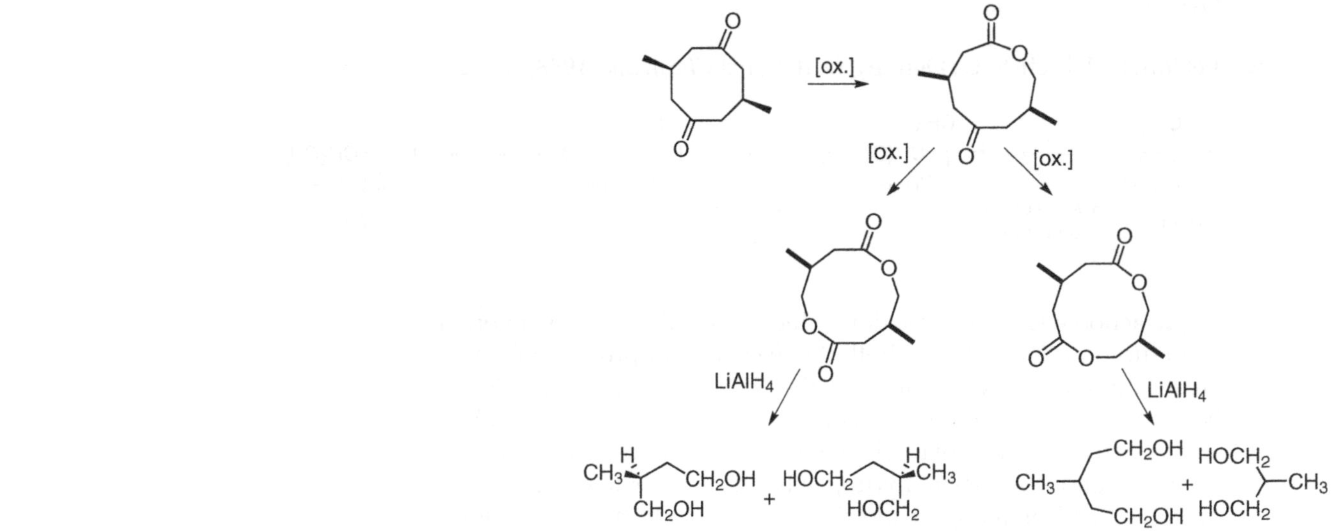

3.32. Jenkins, A. D. *Pure Appl. Chem.* **1981**, *53*, 733.

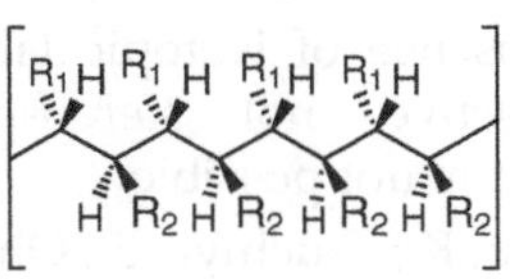

3.33. Structure (a) is chiral; structure (b) is a meso structure because it has a center of inversion.

Roche, A. J.; Duan, J.-X.; Dolbier, W. R., Jr.; et al. *J. Org. Chem.* **2001**, *66*, 7055.

3.34. Carbon 3 in **105** (and also in **106**) is chirotopic because the molecule is chiral. (All atoms and spaces in the molecule are chirotopic.) Carbon 3 is nonstereogenic because interchanging two groups attached to that carbon (e.g. by interchanging the H and OH) does not generate a new stereoisomer; the interchange generates the same stereoisomer (with the Fischer projection turned 180°).

3.35. https://goldbook.iupac.org/terms/view/P04921

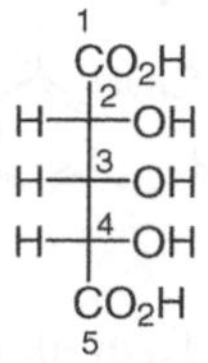

ribaric acid

(2*R*,3*r*,4*S*)-2,3,4-trihydroxypentanedioic acid

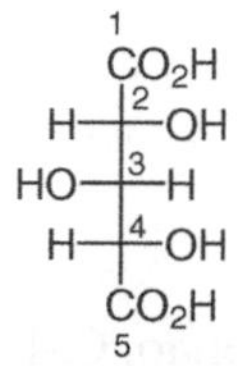

xylaric acid

(2*R*,3*s*,4*S*)-2,3,4-trihydroxypentanedioic acid

3.36. **a.** Takahashi, Y.; Kubota, T.; Fukushi, E.; et al. *Org. Lett.* **2008**, *10*, 3709.

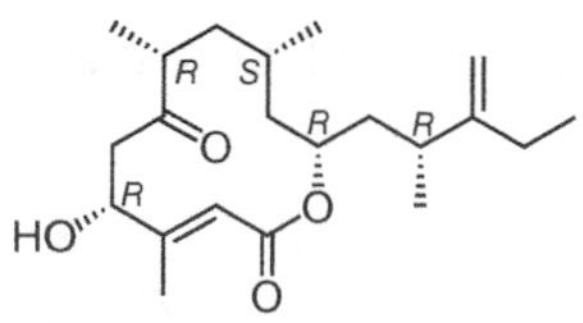

b. El Sayed, K. A.; Laphookhieo, S.; Yousaf, M.; et al. *J. Nat. Prod.* **2008**, *71*, 117. (The atoms are numbered differently in that paper.)

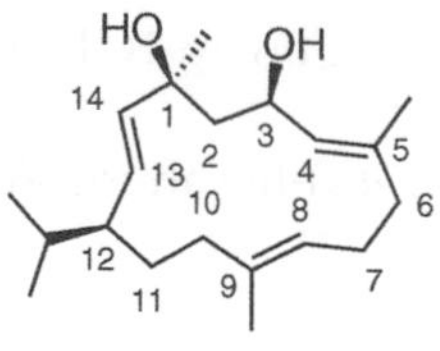

(1*S*,3*R*,4*E*,8*E*,12*S*,13*E*)-12-isopropyl-1,5,9-trimethylcyclotetradeca-4,8,13-triene-1,3-diol

c. Reddy, C. R.; Srikanth, B.; Dilipkumar, U.; et al. *Eur. J. Org. Chem.* **2013**, 525.

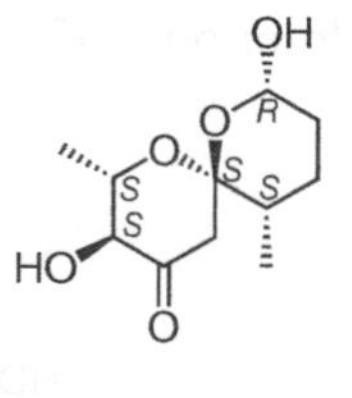

d. Pantin, M.; Brimble, M. A.; Furkert, D. P. *J. Org. Chem.* **2018**, *83*, 7049.

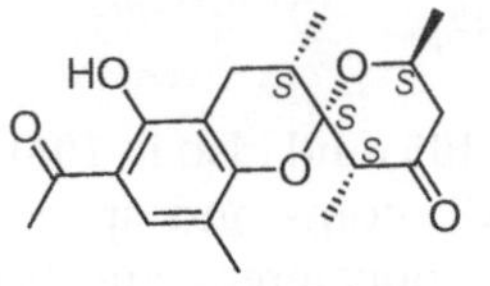

e. Cerulli, A.; Lauro, G.; Masullo, M.; et al. *J. Nat. Prod.* **2017**, *80*, 1703.

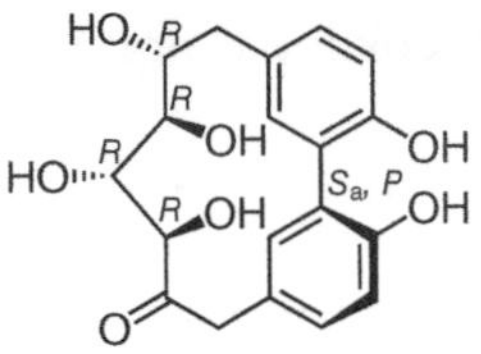

f. Shao, C.-L.; Mou, X.-F.; Cao, F.; et al. *J. Nat. Prod.* **2018**, *81*, 211.

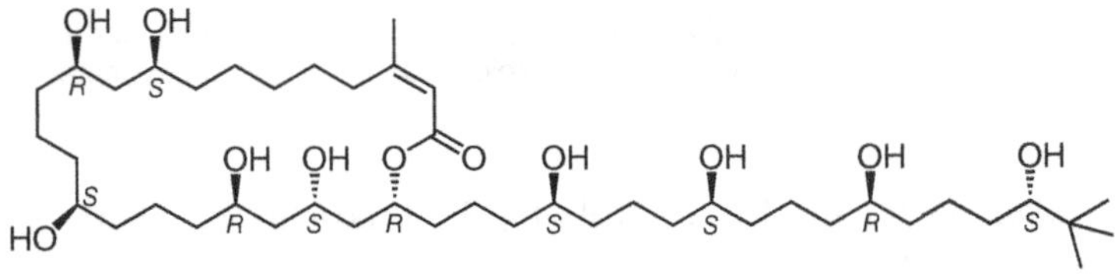

g. Kemkpf, K.; Kempf, O.; Orozco, M.; et al. *J. Org. Chem.* **2017**, *82*, 7791.

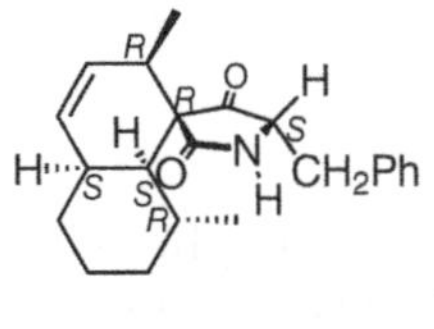

3.37. a. Canoy, W. L.; Cooley, B. E.; Corona, J. A.; et al. *Org. Lett.* **2008**, *10*, 1103.

(3*R*,3*aS*,6*aR*)-hexahydrofuro[2,3-b]furan-3-ol

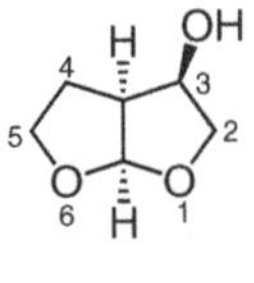

b. Lapointe, G.; Schenk, K.; Renaud, P. *Org. Lett.* **2011**, *13*, 4774.

(3*S*,5*R*,7*aS*,11*aS*)-5-hexyl-3-(hydroxymethyl)octahydro-1H-pyrrolo[2,1-j]quinolin-7(7aH)-one

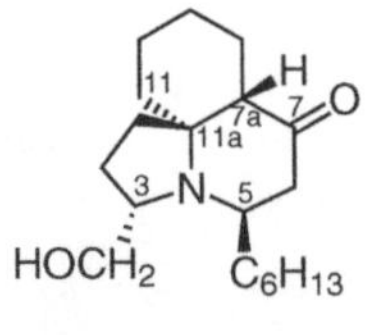

3.38. https://pubchem.ncbi.nlm.nih.gov/compound/Capsidiol

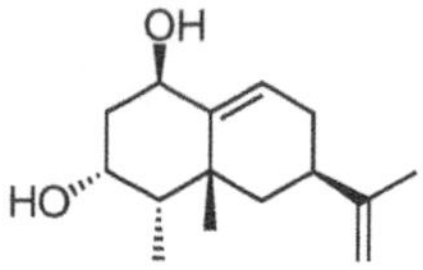

3.39. Baldwin, J. E.; Kostikov, A. P. *J. Org. Chem.* **2010**, *75*, 2767. One 3,6 structure is meso.

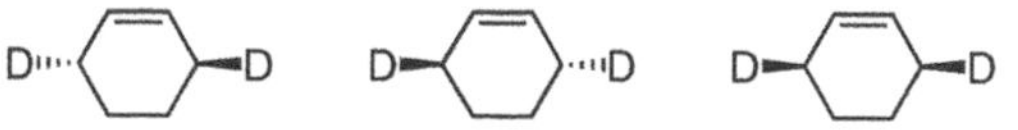

(3*S*,6*S*) (3*R*,6*R*) (3*R*,6*S*) (*meso*)

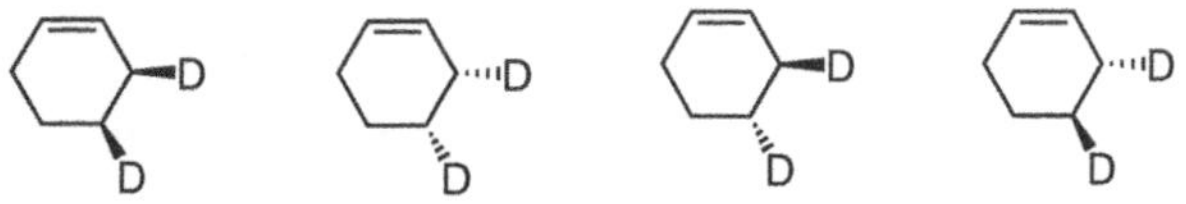

(3*S*,4*S*) (3*R*,4*R*) (3*S*,4*R*) (3*R*,4*S*)

3.40. Mamada, M.; Minamiki, T.; Katagiri, H.; et al. *Org. Lett.* **2012**, *14*, 4062.

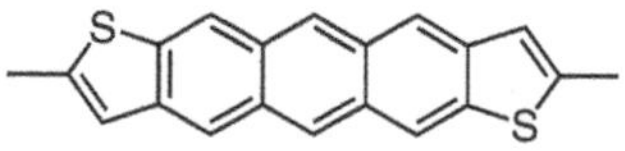

3.41. Kuan, Y. L.; White, J. M. *J. Chem. Soc.* **1994**, 1195.

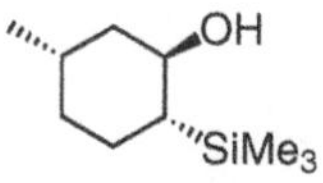

(1*R*,2*R*,5*S*)-5-methyl-2-(trimethylsilyl)cyclohexan-1-ol

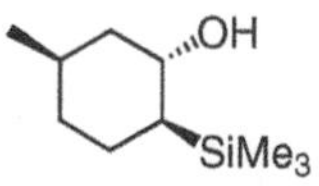

(1*S*,2*S*,5*R*)-5-methyl-2-(trimethylsilyl)cyclohexan-1-ol

CHAPTER 4

Molecular Geometry and Steric Energy

4.1. Ōsawa, E.; Gotō, H.; Oishi, T.; et al. *Pure Appl. Chem.* **1989**, *61*, 597. There are seven single bonds for which rotamers must be considered, so $3^7 = 2187$.

4.2. Allinger, N. L.; Miller, M. A. *J. Am. Chem. Soc.* **1961**, *83*, 2145; Beckett, C. W.; Pitzer, K. S.; Spitzer, R. *J. Am. Chem. Soc.* **1947**, *69*, 2488. The cis isomer has two chair conformations, one in which both methyl groups are axial, and one in which both methyl groups are equatorial. The conformation with two axial methyl groups is 5.5 kcal/mol higher in energy than the e,e conformation. Therefore, almost all molecules of *cis*-1,3-dimethylcyclohexane are in the e,e conformation, which has no butane *gauche* interactions. The trans isomer has one axial and one equatorial methyl group in either of two conformations. Therefore, the trans isomer should be higher in energy (and thus have a higher heat of combustion) than the cis isomer by $2 \times 0.9 = 1.8$ kcal/mol. The experimental value is 1.92 kcal/mol.

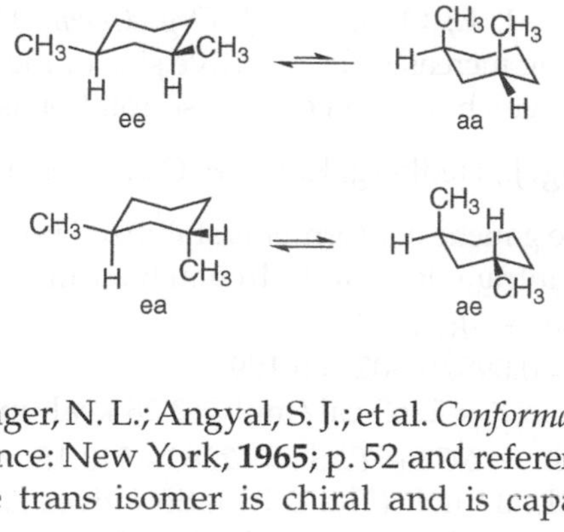

4.3. Eliel, E. L.; Allinger, N. L.; Angyal, S. J.; et al. *Conformational Analysis*; Wiley-Interscience: New York, **1965**; p. 52 and references therein. In both cases, the trans isomer is chiral and is capable of optical

Solutions Manual for Perspectives on Structure and Mechanism in Organic Chemistry, Third Edition.
Felix A. Carroll.

activity. For 1,2-dimethylcyclohexane, the trans isomer can exist predominantly in the diequatorial conformation, while the cis isomer must have one axial methyl group. Therefore, the (chiral) trans isomer is the more stable isomer. In the case of 1,3-dimethylcyclohexane, it is the achiral cis isomer that can have both methyl groups equatorial, so the chiral isomer is the less stable trans isomer.

4.4. Booth, H.; Everett, J. R. *J. Chem. Soc. Perkin Trans. 2* **1980**, 255. Only in the case of *t*-butyl is there a forced enthalpy increase due to van der Waals repulsion; ethyl and isopropyl can adopt conformations in which the extra methyl group(s) point away from the cyclohexane ring; thus, their steric effect is not much greater than that of methyl. For *t*-butyl, however, no such conformation is possible, so the steric preference is much greater.

4.5. Juaristi, E.; Labastida, V.; Antúnez, S. *J. Org. Chem.* **1991**, *56*, 4802. For the conversion of the axial benzyl—equatorial methyl conformer to the axial methyl—equatorial benzyl conformer, ΔS is +1.17 cal/K/mol and $\Delta H = +0.31$ kcal/mol. Since $\Delta G = \Delta H - T\Delta S$, the equilibrium is dominated by the enthalpy term at low temperature, while the entropy term dominates at higher temperature.

4.6. **a.** Hydrogen bonding increases the effective size of the substituent.

b. See Perrin, C. L.; Fabian, M. A. *Abstracts of the 209th National Meeting of the American Chemical Society*, Anaheim, CA, April 2–6, **1995**, Abstract ORGN 6.

The carboxylate ion leads to a structured solvent shell around the ionic carboxylate.

c. Jensen, F. R.; Bushweller, C. H.; Beck, B. H. *J. Am. Chem. Soc.* **1969**, *91*, 344.

Both the bond length and the polarizability of the halogen atom increase along the series F, Cl, Br, I. The two effects reduce nonbonded interactions of an axial halogen substituent with axial halogen atoms as the halogen van der Waals radius increases in the series F, Cl, Br, I.

4.7. Chupp, J. P.; Olin, J. F. *J. Org. Chem.* **1967**, *32*, 2297. Hydrogen bonding increases the effective size of the substituents on the bond about which rotation occurs, so rotation is slowed.

4.8. Huang, J.; Hedberg, K. *J. Am. Chem. Soc.* **1989**, *111*, 6909.

a. The gauche conformer is stabilized both by the gauche effect and by intramolecular hydrogen bonding.

b. $\Delta G° = -RT\ ln\ K$

$K = 0.098/0.902 = 0.109$

$RT \ln K = 2263$ cal/mol $= 2.26$ kcal/mol

c. $\Delta H° = \Delta G° + T\Delta S° = 2263 + 513.15 \times 0.81 = 2678$ cal/mol $=$ 2.7 kcal/mol. (The calculation gives $\Delta U°$ (the internal energy), where $\Delta U° = \Delta H° + \Delta PV$, but the ΔPV term should be near 0 for interconversion of conformers.)

4.9. For *n*-heptane: 2 × (−10.08) + 5 × (−4.95) = −44.91. There are no butane gauche interactions. The literature value is −44.88 kcal/mol.[1]

For 2-methylhexane: 3 × (−10.08) + 3 × (−4.95) + (1.90) = −46.99 kcal/mol. There is one butane gauche interaction, so the corrected value is −46.99 + 0.8 = −46.19 kcal/mol. The experimental value is −46.59 kcal/mol.

For 3-methylhexane, the count is as above, but there are 2 butane gauche interactions, giving a net −45.39 kcal/mol. The literature value is −45.96 kcal/mol.

For 2,2-dimethylpentane, the count is 4 × (−10.08) + 2 × (−4.95) + 1 × (0.5) = −49.72 kcal/mol.

There are 2 butane gauche interactions, so 1.6 kcal/mol is added, giving a total of −48.12 kcal/mol. The literature value is −49.27 kcal/mol.

For 2,3-dimethylpentane, the count is 4 × (−10.08) + 1 × (−4.95) + 2 × (−1.90) = −49.07. Now add 2.4 kcal/mol for 3 butane gauche interactions to get −46.67 kcal/mol. The literature value is −47.62.

For 3,3-dimethylpentane, the count is the same as for 2,2-dimethylpentane. Thus, the strain-free value is −49.72 kcal/mol. There are 4 butane gauche interactions, so +3.2 kcal/mol is added to get −46.52 kcal/mol. The literature value is −48.17 kcal/mol.

Calculations for the other three isomers of C_7H_{16} can be carried out in similar fashion:

For 3-ethylpentane: 3 × (−10.08) + 3 × (−4.95) + 1 × (−1.90) = −46.99 kcal/mol. There are 3 butane gauche interactions, so the net is −44.59 kcal/mol. Literature is −45.33 kcal/mol.

For 2,4-dimethylpentane, 4 × (−10.08) + 1 × (−4.95) + 2 × (−1.90) = −49.07 kcal/mol. There are 2 butane gauche interactions, so the net is −47.47 kcal/mol. The literature value is −48.28 kcal/mol.

For 2,2,3-trimethylbutane, 5 × (−10.08) + 1 × (−1.9) + 1 × (0.5) = −51.8 kcal/mol. Four butane gauche interactions add 3.2, so the net is −48.6 kcal/mol. The literature value is −48.95 kcal/mol.

These data suggest that the heats of formations of a series of isomeric alkanes are influenced by the number of substituents (branches), the steric size of the substituents, and the proximity of the substituents to each other along the chain. Increasing the number of branches lowers the heat of formation, as does increasing the size of the substituents on isomers with the same number of substituents. Substituents on adjacent carbon atoms interfere sterically more than do substituents on the same carbon atom, leading to a less negative heat of formation.

4.10. Twistane or tricyclo[4.4.0.0^{3,8}]decane: Whitlock, H. W., Jr. *J. Am. Chem. Soc.* **1962**, *84*, 3412. See also Olbrich, M.; Mayer, P.; Trauner, D. *Org. Biomol. Chem.* **2014**, *12*, 108.

[1] Stull, D. R.; Westrum, E. F., Jr.; Sinke, G. C. *The Chemical Thermodynamics of Organic Compounds*; John Wiley & Sons: New York, 1969; pp. 249–252; also see Cox, J. D.; Pilcher, G. *Thermochemistry of Organic and Organometallic Compounds*; Academic Press: New York, 1970; p. 157 and references therein.

Basketane or pentacyclo[4.4.0.0^{2,5}.0^{3,8}.0^{4,7}]decane: Gassman, P. G.; Yamaguchi, R. *J. Org. Chem.* **1978**, *43*, 4654. See also *J. Comput. Chem. Jpn. Int. Ed.* **2015**, *1*, 1.

Tricyclo[2.1.0.0^{1,3}]pentane: Wiberg, K. B.; McMurdie, N.; McClusky, J. V.; et al. *J. Am. Chem. Soc.* **1993**, *115*, 10653.

4.11. Maier, G. *Angew. Chem. Int. Ed. Engl.* **1988**, *27*, 309. The *t*-butyl groups block the attack of reagents other than a proton on the tetrahedrane skeleton. In addition, they provide a what has been called a "corset effect," which holds the molecule in a tetrahedral shape. Any reaction that involves distortion of the tetrahedrane framework would increase the steric barrier of the bulky substituents.

4.12.

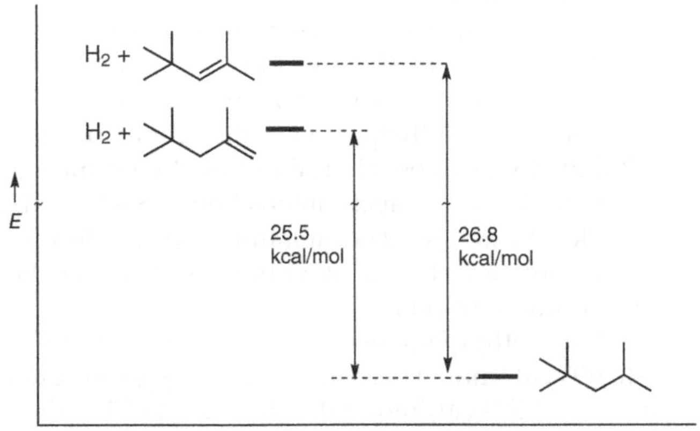

H_2 +

H_2 +

E

25.5 kcal/mol

26.8 kcal/mol

2,4,4-Trimethyl-1-pentene is less substituted. However, 2,4,4-trimethyl-2-pentene has a *t*-butyl group that is cis to a methyl group, which raises the energy of the compound due to van der Waals repulsion. The experimental data were reported by Turner, R. B.; Nettleton, D. E., Jr.; Perelman, M. *J. Am. Chem. Soc.* **1958**, *80*, 1430. Brown, H. C.; Berneis, H. L. *J. Am. Chem. Soc.* **1953**, *75*, 10 had proposed the steric explanation for the greater stability of the less substituted isomer. For a discussion, see Saunders, W. H., Jr.; Cockerill, A. F. *Mechanisms of Elimination Reactions*; Wiley-Interscience: New York, **1973**; p. 173.

4.13. Golan, O.; Goren, Z.; Biali, S. E. *J. Am. Chem. Soc.* **1990**, *112*, 9300; Juaristi, E.; Labastida, V.; Antúnez, S. *J. Org. Chem.* **1991**, *56*, 4802. (a) is *ap*; (b) is +*sc*; (c) is –*sc*; (d) is –*sc*. In (a), the two groups that determine the conformational designation are the CH_3 group on the upper carbon atom and the hydrogen of the lower carbon atom because in each case these substituents are different from the other two substituents on the carbon atoms to which they are attached. In (b) the two determining substituents are both hydrogens, for the same reasons. In (c) and (d), the two groups are the hydrogen atom on the cyclohexane ring and the phenyl group on the $CH_2C_6H_5$ substituent.

4.14. The *sc* conformer is more stable because it keeps the two *t*-butyl groups away from the isopropyl group on the aromatic ring. Casarini, D.; Coluccini, C.; Lunazzi, L.; et al. *J. Org. Chem.* **2005**, *70*, 5098.

4.15. Barton, D. H. R.; Cookson, R. C. *Quart. Rev. Chem. Soc.* **1956**, *10*, 44; Johnson, W. S. *J. Am. Chem. Soc.* **1953**, *75*, 1498. The trans isomer is more stable by 2.4 kcal/mol because of three butane gauche interactions in the cis isomer.

4.16. Barton, D. H. R.; Cookson, R. C. *Quart. Rev. Chem. Soc.* **1956**, *10*, 44; Johnson, W. S. *J. Am. Chem. Soc.* **1953**, *75*, 1498. The *trans-anti-trans* isomer can have all chair conformations (with one gauche interaction), but the *trans-syn-trans* isomer has one cyclohexane ring in a boat conformation. Therefore, the former isomer is more stable by at least 5.5 kcal/mol. Other steric interactions are also present in the *trans-syn-trans* isomer. A molecular mechanics calculation indicated that the *trans-syn-cis* isomer is more stable than the *trans-syn-trans* isomer by about 7.5 kcal/mol.

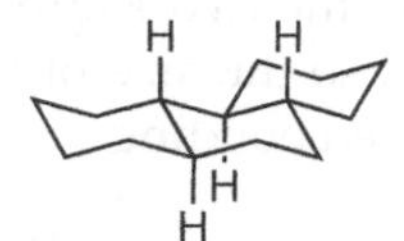

trans-anti-trans

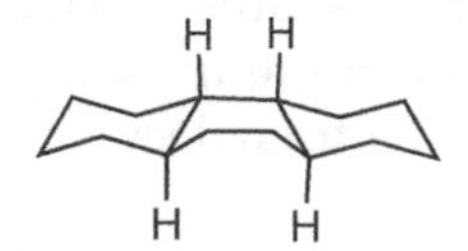

trans-syn-trans

4.17. Jensen, F. R.; Bushweller, C. H.; Beck, B. H. *J. Am. Chem. Soc.* **1969**, *91*, 344.

$$\Delta G^\circ = -RT \ln K.$$

In these calculations, it is important to convert ΔG° values in kcal/mol to values in cal/mol because the value of R often used (1.987 cal/K/mol) is in cal, not kcal. In addition, temperature must be converted to degrees K. The calculation yields $K = 6.95$. Thus, the distribution of equatorial and axial conformers is 87.4% and 12.6%, respectively.

4.18. Jensen, F. R.; Bushweller, C. H.; Beck, B. H. *J. Am. Chem. Soc.* **1969**, *91*, 344.

$$\Delta G^\circ = 0.50 \text{ kcal / mol.}$$

4.19. Booth, H.; Everett, J. R. *J. Chem. Soc. Perkin Trans. 2* **1980**, 255. The entropy term dominates ΔG at 300 K; at 40 K the ΔH term dominates.

4.20. Eliel, E. L.; Manoharan, M. *J. Org. Chem. 1981*, *46*, **1959**.

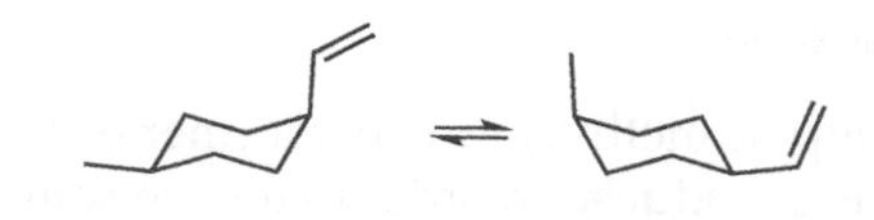

$$\Delta G^\circ = A_{\text{methyl}} - A_{\text{vinyl}} = 0.06 \text{ kcal / mol}$$

Using 1.74 as the A value for the methyl group gives an A value for the vinyl group of 1.68.

4.21. The A value is 1.19 at 298 K. The A value is 1.14 at 178 K, so the equatorial conformer is favored to a lesser extent at that temperature. Juaristi, E.; Labastida, V.; Antúnez, S. *J. Org. Chem.* **2000**, *65*, 969.

4.22. The chair conformation is the global minimum on the surface shown, while the half-chair conformation is the global maximum. (It would be reasonable to state that a global maximum is also a local maximum and that a global minimum is also a local minimum since local maxima and minima are defined with regard to only a small portion of the potential energy surface without regard to the rest of the surface.) The twist boat conformations are local minima, while the boat conformation is a local maximum.

4.23. Kuhn, L. P. *J. Am. Chem. Soc.* **1958**, *80*, 5950. The two OH groups must be gauche in order for intramolecular hydrogen bonding to be observed. In the meso diastereomer, the two R groups must also be gauche if the OH groups are gauche, so steric repulsion of the *t*-butyl groups makes the energy of this conformation prohibitive. In the racemic diastereomer, the two R groups are anti when the two OH groups are gauche, so the size of the R groups does not affect intramolecular hydrogen bonding.

OH
H OH
H R
R
meso

OH
R OH
H R
H
(+) or (–)

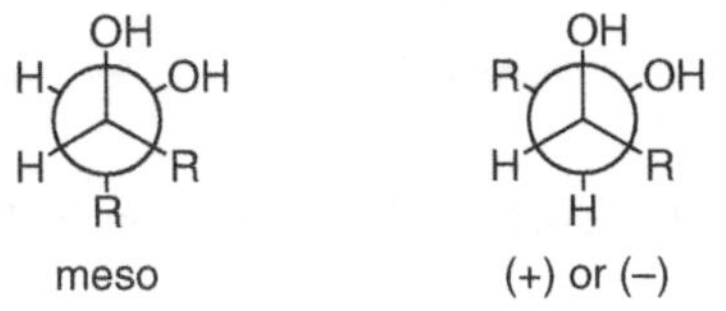

4.24. The figure on the left is for the meso structure because it is symmetric.

4.25. **a.** Conformation B is puckered cyclobutane.

b. In comparing the planar conformation to the puckered conformation, the major differences are significantly greater angle strain but even more significantly less torsional strain in the puckered conformation. There is slightly greater van der Waals strain in the puckered conformer. There is also somewhat greater angle strain in the puckered conformer, but this is offset somewhat by a smaller stretch-bend term. Overall, the puckered conformation is more stable than the planar conformation by 0.92 kcal/mol.

4.26. Conformer B has a boat conformation for the six-membered ring, while A has a chair conformation. The overall difference in steric energy is similar to the twist-boat—chair energy difference in cyclohexane.

4.27. The flagpole methyl group conformer is A because greater torsional strain is introduced in order to relieve some van der Waals strain of the flagpole methyl with the flagpole hydrogen near it.

4.28. A is *aa*. B is g^+a. C is g^+g^-. D is g^-g^-. For a related paper, see Klauda, J. B.; Brooks, B. R.; MacKerell, A. D., Jr.; et al. *J. Phys. Chem. B.* **2005**, *109*, 5300.

4.29. 65.4%, 31.6%, and 3.1%, respectively. Wang, F.; Polavarapu, P. L. *J. Phys. Chem. A* **2000**, *104*, 6189.

4.30. Using the relationship $J = 2.15 \times$ (mole fraction chair conformation) $+ 11.17 \times$ (mole fraction boat conformation) $= 3.90$ Hz, the mole fraction of chair conformations is determined to be 0.81.

The reported value is 0.82. Jaime, C.; Ōsawa, E.; Takeuchi, Y.; et al. *J. Org. Chem.* **1983**, *48*, 4514.

4.31. Grein, F. *J. Phys. Chem. A* **2002**, *106*, 3823. See also Leroux, F. *ChemBioChem* **2004**, *5*, 644. The order is biphenyl < 2-fluorobiphenyl < 2-chlorobiphenyl < 2-bromobiphenyl, which is based on substituent size.

4.32. Another source of strain not shown in that figure is torsional strain. The drawing here shows a portion of the minimum energy (DFT) conformation of *trans*-cycloheptene. The dihedral angle for H(C2)–C2–C3–H(C3) is calculated to be 30°.

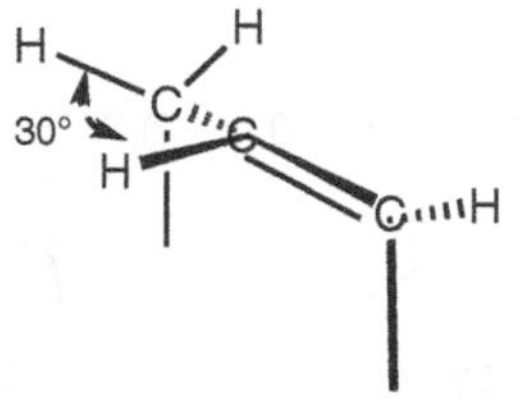

4.33. Wang, X.; Lau, K.-C.; Li, W.-K. *J. Phys. Chem. A* **2009**, 3413. Helvetane features a prismane-like array of cyclobutane rings arranged in the geometry of a cross, which suggests the national flag of Switzerland. The four-membered rings have considerable angle strain from the 90° internuclear bond angles and torsional strain from the eclipsed bonds.

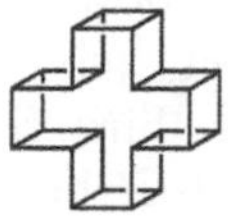

Fokin, A. A.; Schreiner, P. R. in Dodziuk, H., Ed. *Strained Hydrocarbons*; Wiley-VCH: Weinheim, Germany, **2009**; p. 17. Bowlane features a pyramidal carbon at one apex of the structure.

4.34. Wang, F.; Polavarapu, P. L.; Lebon, F.; et al. *J. Phys. Chem. A* **2002**, *106*, 12365. The source uses T instead of A and t instead of a.

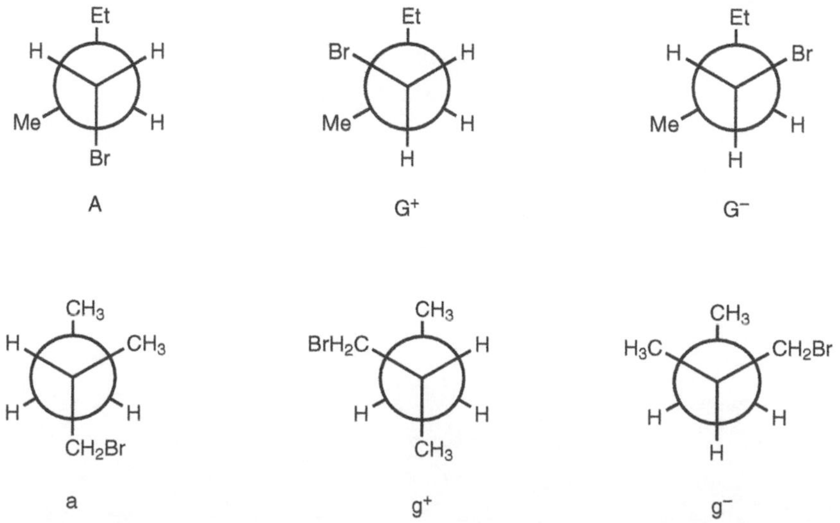

4.35. Ingavat, N.; Mahidol, C.; Ruchirawat, S.; et al. *J. Nat. Prod.* **2011**, *74*, 1650. [5.5.5.6]fenestrane

4.36. Using the 0° C1–C2–C3–C4 dihedral angle implied by the line formula for structure **82** as the initial geometry for a DFT geometry optimization calculation results in an energy minimum with a −1.8° dihedral angle for the C1–C2–C3–C4 bonds. However, an initial geometry closer to 90° results in a minimum (**82a**) with an 88.6° dihedral angle that is 2.93 kcal/mol lower in energy. This example serves as a reminder that relative energy comparisons can be misleading unless care is taken to find the global minimum of each structure and not just a local minimum based on a convenient input geometry.

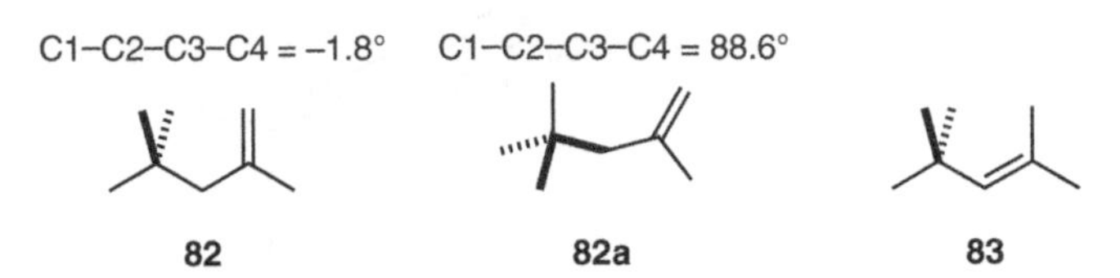

4.37. Carroll, F. A.; Blauch, D. N. *Chem. Educator* **2016**, *21*, 162. In the first figure, the two horizonal axes represent the dihedral angles for rotation about the C1–C2 and the C3–C4 bonds of butane. The second figure represents rotation about the C1–C2 and C2–C3 bonds of butane. All of the peaks are higher than those in the first figure because they show not only torsional strain but also steric strain from $H-CH_3$ (six peaks) or CH_3-CH_3 (three peaks) interactions. In the third figure, one horizontal axis represents rotation about the C2–C3 bond of butane, while the other horizontal axis represents variation in the C1–C2–C3 bond angle.

CHAPTER 5

Reactive Intermediates

5.1. For a discussion, see Wentrup, C. *Reactive Molecules;* John Wiley & Sons: New York, 1984; pp. 33–34, which cites the original literature for these compounds. Structure **93** has a stable (singlet) resonance structure with no radical character, so it exhibits predominantly singlet character. Structure **94** cannot generate a Kekulé resonance structure (having only double and single bonds) that does not have radical character. Thus, it is a triplet species because triplet diradicals typically are lower in energy than are singlet diradicals (see Chapter 12).

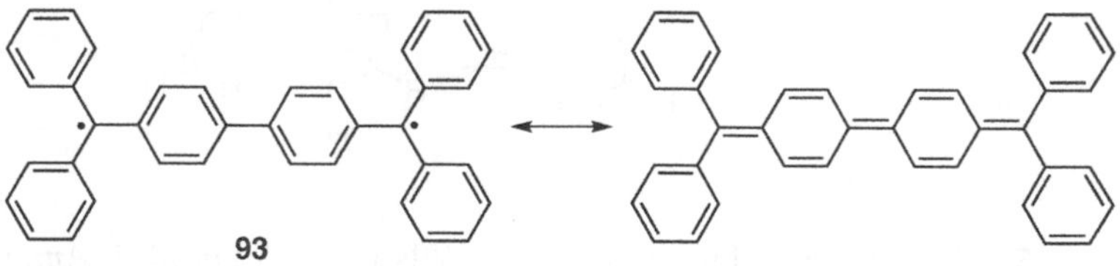

93

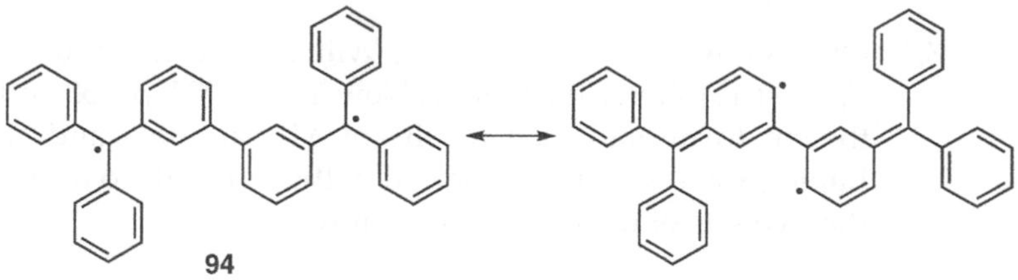

94

Solutions Manual for Perspectives on Structure and Mechanism in Organic Chemistry, Third Edition.
Felix A. Carroll.

5.2. Ayscough, P. B. *Electron Spin Resonance in Chemistry*; Methuen & Co.: London, 1967; p. 298; Bunce, N. J. *J. Chem. Educ.* **1987**, *64*, 907 (especially p. 910). The electron paramagnetic resonance spectrum of the benzyl radical has been interpreted to mean that about 50% of the unpaired electron density is on the benzylic carbon atom, while 15.8% is on each of the two ortho carbon atoms and 18.6% is on the para carbon atom. See also the discussion by Fleming, I. *Frontier Orbitals and Organic Chemical Reactions*; Wiley-Interscience: London, 1976; p. 60.

5.3. Olah, G. A. *Carbocations and Electrophilic Reactions*; John Wiley & Sons: New York, 1974; p. 20. Using the formula

$$J_{^{13}C\text{–}H} = \frac{500}{(1+\lambda^2)} = 169 \text{ Hz},$$

gives $(1 + \lambda^2) = 2.958$. Then percent *s* character is

$$\frac{100\%}{(1+\lambda^2)} = 33.8\%$$

Therefore, the carbon orbital is slightly more *s*-like than is an sp^2 hybrid orbital.

5.4. Olah, G. A. *Carbocations and Electrophilic Reactions*; John Wiley & Sons: New York, 1974; p. 63; Olah, G. A.; White, A. M. *J. Am. Chem. Soc.* **1967**, *89*, 3591.

5.5. Olah, G. A.; Mateescu, G. D.; Wilson, L. A.; et al. *J. Am. Chem. Soc.* **1970**, *92*, 7231. Resonance delocalization stabilizes the charge, making all of the trityl carbons have a similar electronic environment.

5.6. Olah, G. A.; Prakash, G. K. S.; Williams, R. E.; et al. *Hypercarbon Chemistry*; John Wiley & Sons: New York, 1987; p. 153 and reference therein to unpublished results of M. Saunders and co-workers. Rapid proton shift from one end of the molecule to the other makes the two sets of methyl protons equivalent.

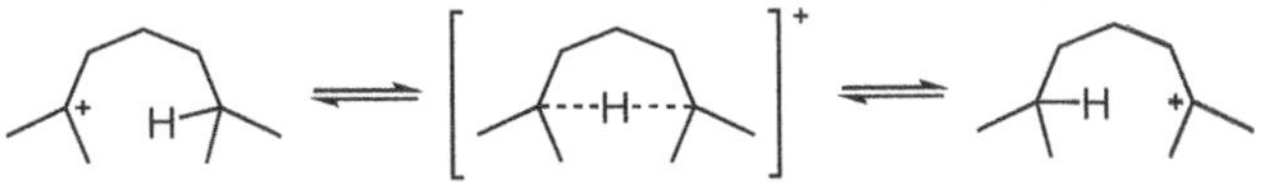

5.7. Olah, G. A.; Prakash, G. K. S.; Sommer, J. *Superacids*; Wiley-Interscience: New York, 1985; p. 84. The authors suggested that the isomerization involves a primary carbocation, then a protonated cyclopropane, then a linear secondary carbocation. The latter can go back to a *t*-butyl cation with a change in the position of the carbon atoms. Similar rearrangements were discussed by Saunders, M.; Vogel, P.; Hagen, E. L.; et al. *Acc. Chem. Res.* **1973**, *6*, 53.

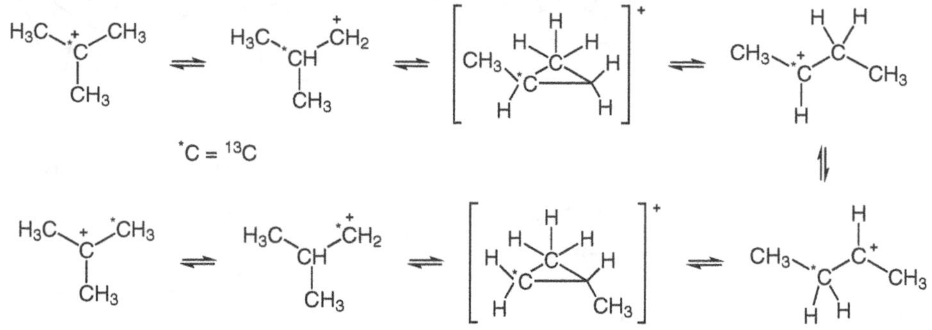

*C = ^{13}C

The activation energy is consistent with this mechanism.

5.8. Walborsky, H. M.; Periasamy, M. P. *J. Am. Chem. Soc.* **1974**, *96*, 3711.

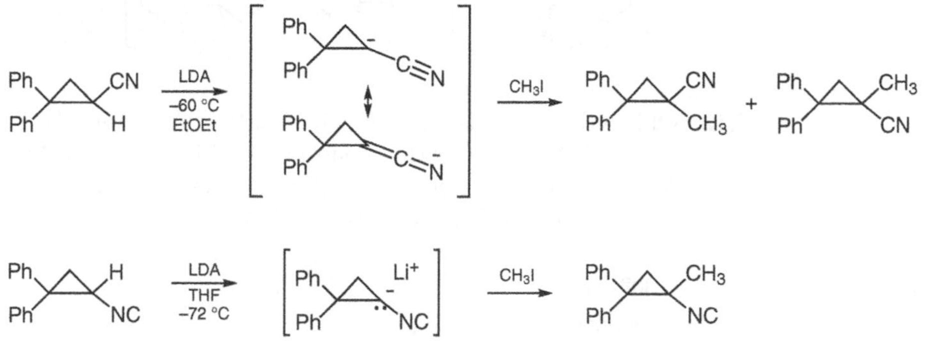

The cyano group is able to stabilize and flatten the carbanion by resonance. However, the isocyano group cannot form a double bond with the ring, because to do so would involve the following type of resonance:

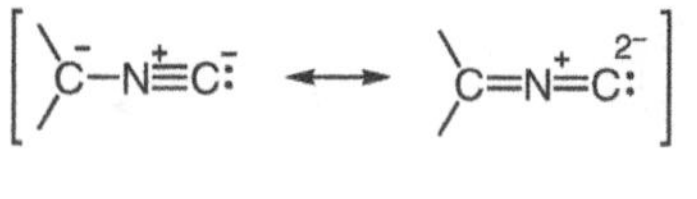

5.9. Zimmerman, H. E.; Zweig, A. *J. Am. Chem. Soc.* **1961**, *83*, 1196. Intramolecular reaction converts the alkyllithium to an aryllithium, which then reacts with CO_2 to give the benzoic acid derivative.

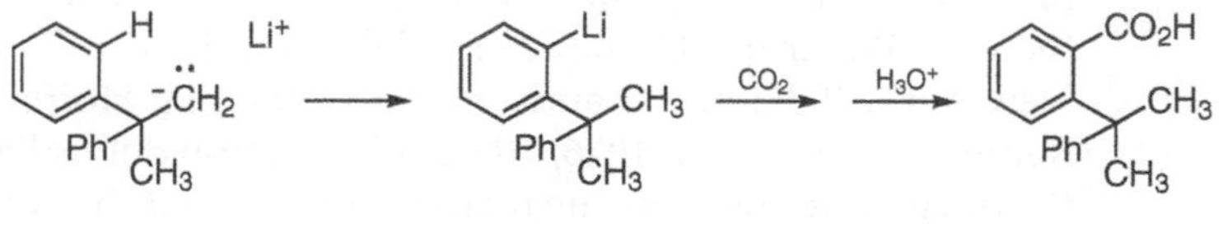

5.10. $\Delta H°$ is 11 kJ/mol because the same radical is produced from both bond dissociations. Agapito, F.; Nunes, P. M.; Costa Cabral, B. J.; et al. *J. Org. Chem.* **2007**, *72*, 8770.

5.11. Fitjer, L.; Quabeck, U. *Angew. Chem. Int. Ed. Engl.* **1987**, *26*, 1023.

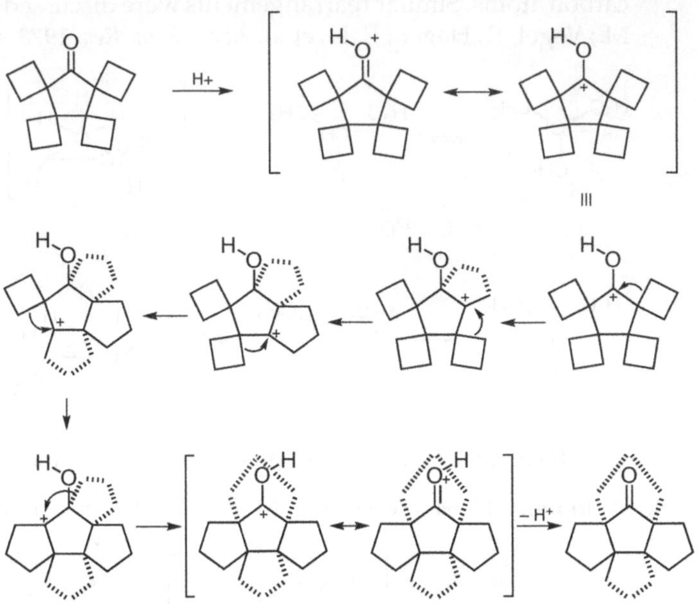

5.12. Oldroyd, D. M.; Fisher, G. S.; Goldblatt, L. A. *J. Am. Chem. Soc.* **1950**, *72*, 2407. Also see the discussion by Walling, C. in de Mayo, P., Ed. *Molecular Rearrangements*, Part I; Wiley-Interscience: New York, 1963; pp. 407–455 (particularly p. 440).

Initiation

$R^{\bullet} + CCl_4 \rightarrow RCl + {}^{\bullet}CCl_3$

Propagation

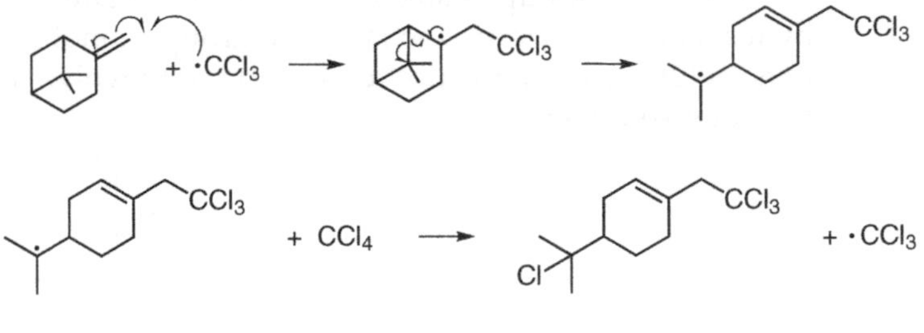

5.13. For a discussion, see Moss, R. A.; Jones, M., Jr. in Jones, M., Jr.; Moss, R. A., Eds. *Reactive Intermediates*, Vol. 1; Wiley-Interscience: New York, 1978; pp. 67–116 (especially p. 97) and references therein; Wentrup, C. *Reactive Molecules*; John Wiley & Sons: New York, 1984; p. 238 and references therein; Hoffmann, R. W.; Reiffen, M. *Chem. Ber.* **1976**, *109*, 2565. Electron delocalization from the oxygens to the carbenic *p* orbital means that the carbene is not as electrophilic as most simple alkenes and thus is less likely to react with nucleophiles.

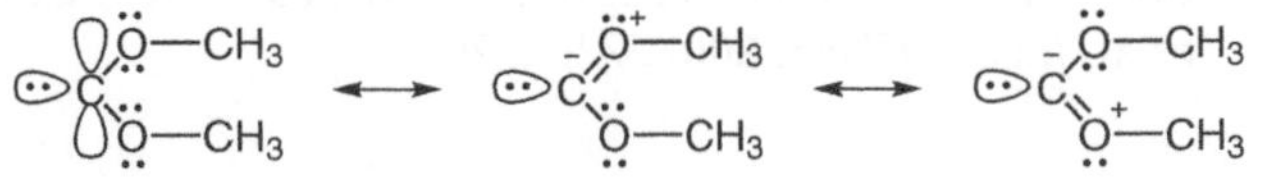

5.14. Perkins, M. J. *Adv. Phys. Org. Chem.* **1980**, *17*, 1.

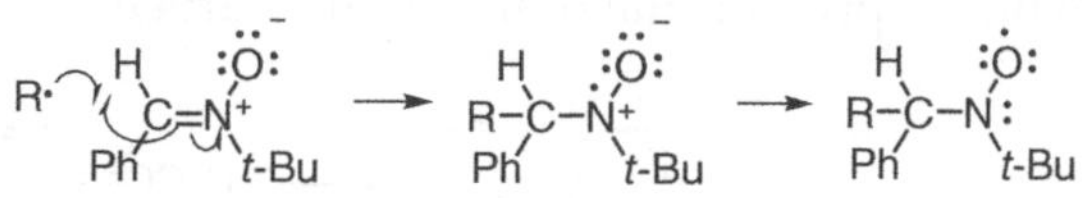

5.15. Friedman, L.; Shechter, H. *J. Am. Chem. Soc.* **1961**, *83*, 3159. The first two products are formed by insertion of the carbene into a transannular carbon–hydrogen bond. The last two products are formed by hydrogen shift.

5.16. Jones, R. R.; Bergman, R. G. *J. Am. Chem. Soc.* **1972**, *94*, 660. The cyclization of a 1,5-hexadiyn-3-ene to a 1,4-dehydrobenzene is known as the Bergman cyclization.

a. Five other structures are possible. Four of them (as well as **97**) have both a C_2 axis and a plane of symmetry. However, structure (v) below has only a C_2 axis.

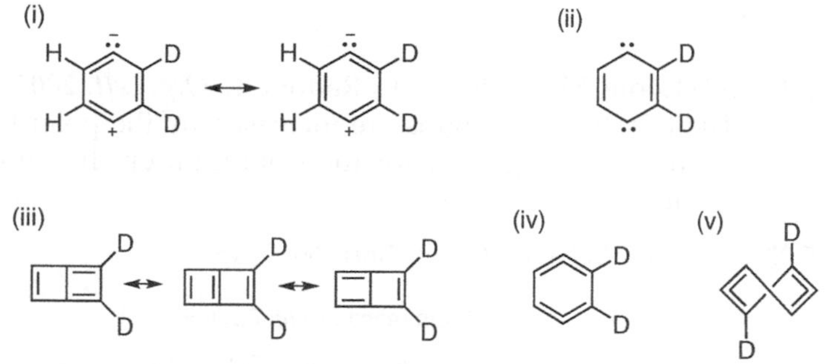

(i) (ii)

(iii) (iv) (v)

b. The formation of benzene is most readily explained as being the product of radical atom abstraction and coupling reactions. Intermediates with ionic character would be expected to react with methanol to yield anisole, and those with allenic character would not readily abstract hydrogen atoms from alkane solvent molecules.

5.17. Closure (a) is 5-exo-dig; closure (b) is 6-endo-dig. For a discussion, see Alabugin, I. V.; Manoharan, M. *J. Am. Chem. Soc.* **2005**, *127*, 9534. Closure (c) is 6-endo-trig: Liu, F.; Liu, K.; Yuan, X.; et al. *J. Org. Chem.* **2007**, *72*, 10231.

5.18. Geometric constraints prevent the cyclopropyl radical center from being planar, but the cyclohexyl radical can achieve a more nearly planar geometry at the radical center. The cyclohexyl radical is therefore in an orbital with significant *p* character. For a discussion, see Fessenden, R. W.; Schuler, R. H. *J. Chem. Phys.* **1963**, *39*, 2147. For calculated structures of the radicals, see Bera, P. P.; Horný, L.; Schaefer, H. F., III *J. Am. Chem. Soc.* **2004**, *126*, 6692.

5.19. Singlet carbenes generally prefer smaller bond angles to the carbenic center than do triplet carbenes. As the ring size increases, therefore, the singlet becomes destabilized in comparison with the triplet. For a computational study, see Nicolaides, A.; Matushita, T.; Tomioka, H. *J. Org. Chem.* **1999**, *64*, 3299.

5.20. The para isomer is the singlet due to resonance donation of electrons to the empty *p* orbital of the singlet carbene:

This interaction is not possible with the meta isomer. Song, M.-G.; Sheridan, R. S. *32nd Reaction Mechanisms Conference*, Chapel Hill, NC, June 25–28, 2008, Poster P-45.

5.21. Clive, D. L. J.; Sunasee, R. *Org. Lett.* **2007**, *9*, 2677. The product is formed by a radical chain pathway.

5.22. See Coote, M. L.; Pross, A.; Radom, L. *Org. Lett.* **2003**, *5*, 4689. Yes. The C–F bond strengths are increased by the polarity of the C–F bond, which is greater for the *t*-butyl group than for the methyl group.

5.23. Olah, G. A. *J. Org. Chem.* **2001**, *66*, 5943.

From reaction with FSO_3H – SbF_5

From reaction with SbF_5 / SO_2

From reaction with HF / SbF_5 / SO_2

5.24. The other product is (triplet) diadamantylcarbene. Prakash, G. K. S. *J. Org. Chem.* **2006**, *71*, 3661.

Ad = 1-adamantyl

5.25. The empty *p* orbital on C2 of the 2-adamantyl carbocation is oriented 90° with respect to the C–H bond on C1. For a discussion, see Schleyer, P. v. R.; Lam, L. K. M.; Raber, D. J.; et al. *J. Am. Chem. Soc.* **1970**, *92*, 5246.

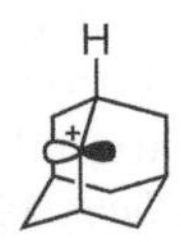

5.26. The hyperconjugative interaction that stabilizes the carbocation decreases the bonding between C_β and C_γ and puts some positive charge on C_γ (but not on C_β). See Prakash, G. K. S.; Schleyer, P. v. R., Eds. *Stable Carbocation Chemistry*; John Wiley & Sons: New York, **1997**; p. 149.

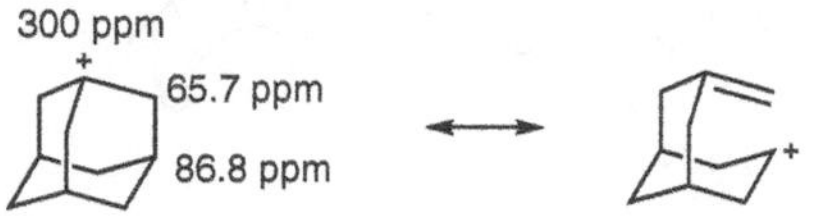

5.27. The stability varies as the natural logarithm of the number of atoms because a larger structure can more effectively stabilize a carbocation by dispersion. The correlation is HIA = −7.382 ln(n) + 276.13, with R^2 = 0.998. Lossing, F. P.; Holmes, J. L. *J. Am. Chem. Soc.* **1984**, *106*, 6917 and references therein.

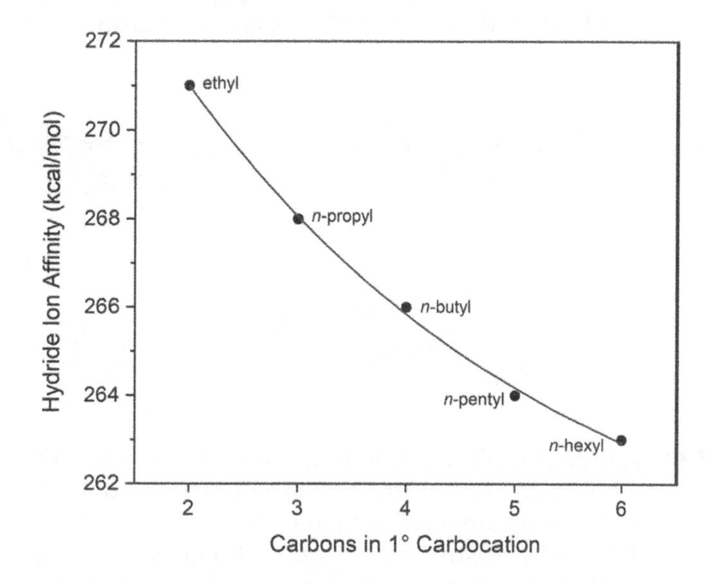

5.28. There is greater strain in the H-bridged transition structures for the smaller rings. See Chandrasekhar, J.; Schleyer, P. v. R. *Tetrahedron Lett.* **1979**, 4057.

5.29. Brinker, U. H.; Weber, J. *Angew. Chem. Int. Ed. Engl.* **1997**, *36*, 1623.

a. The mechanism is illustrated below for the methoxy-substituted reactant.

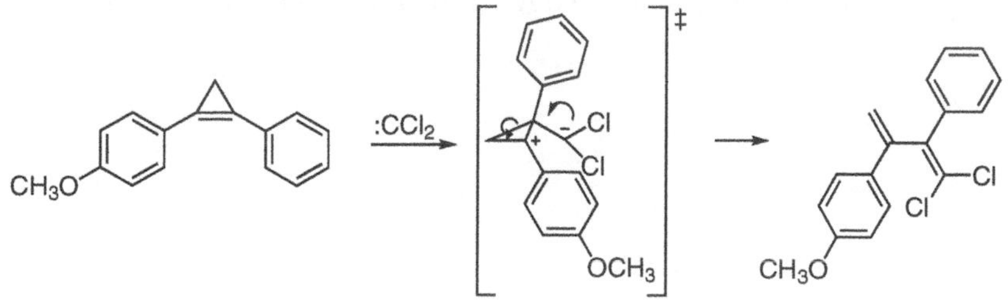

b. As shown above, the preferred pathway has the benzylic carbocation on the *p*-methoxy-substituted aromatic ring.

5.30. Nickon, A.; Lambert, J. L. *J. Am. Chem. Soc.* **1962**, *84*, 4604. See the mechanism in Gassman, P. G.; Zalar, F. V. *J. Am. Chem. Soc.* **1966**, *88*, 3070.

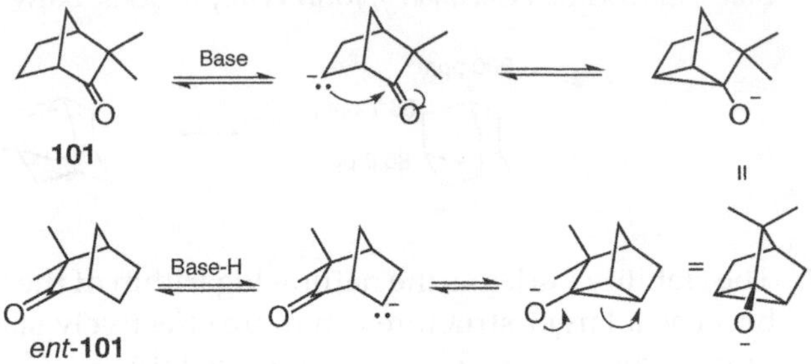

5.31. Eom, D.; Park, S.; Park, Y.; et al. *Org. Lett.* **2012**, *14*, 5392. See also Nardella, R. R.; Klumpp, D. A. *Chem. Rev.* **2013**, *113*, 6905.

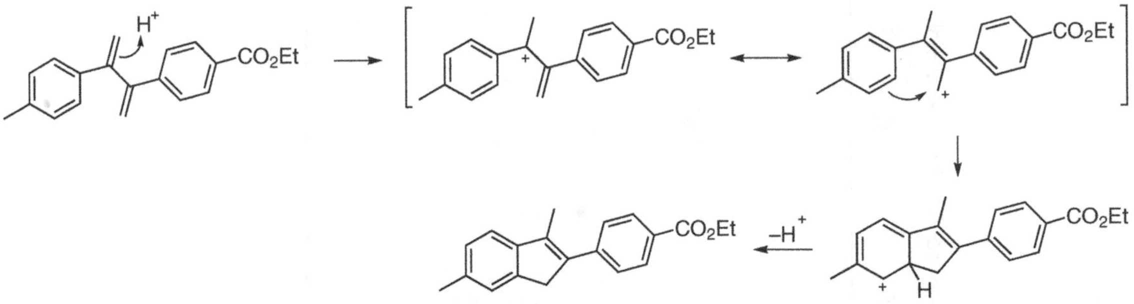

5.32. A radical chain reaction initiated by the reaction of triethylborane with air involves abstraction of a hydrogen α to the N because those C–H bonds are weaker than C–H bonds α to oxygen. (See data in Table 5.1.) Abstraction of a C–H bond on the ring instead of the methyl group is more favorable because of hyperconjugative stabilization of the radical center. See Yoshimitsu, T.; Arano, Y.; Nagaoka, H. *J. Am. Chem. Soc.* **2005**, *127*, 11610.

5.33. Inconsistencies between expected and observed experimental results usually prompt investigators to examine first the validity of their experimental data, next the design of their experiments, and only then the assumptions on which the experiments are based. It seems unnecessary routinely to repeat prior experiments unless such inconsistencies suggest the likelihood of error in earlier work.

CHAPTER 6

Determining Reaction Mechanisms

6.1. The recombination of the fragments of dissociation of the molozonide might be occurring within a solvent cage, but a very low concentration of alkene could have produced the same results. In the latter case, increasing the concentration of the reactant should increase the chance of finding cross ozonolysis products. Indeed, carrying out the ozonolysis of a mixture of 3-hexene and 4-octene produced 3-heptene ozonide. See Murray, R. W.; Williams, G. J. *Adv. Chem. Ser.* **1968**, *77*, 32; Murray, R. W. *Acc. Chem. Res.* **1968**, *1*, 313 and references therein. See also Murray, R. W.; Story, P. R.; Loan, L. D., Jr. *J. Am. Chem. Soc.* **1965**, *87*, 3025.

6.2. Carry out the reaction in presence of DCl. Observing deuterated 2-chloropropane would support the first mechanism. Failure to observe deuterated 2-chloropropane would be consistent with (but not "prove") the second mechanism. See Nash, L. M.; Taylor, T. I.; Doering, W. v. E. *J. Am. Chem. Soc.* **1949**, *71*, 1516.

6.3. Long, F. A.; Pritchard, J. G. *J. Am. Chem. Soc.* **1956**, *78*, 2663.

Both reactions are highly but not completely regioselective. In each case, most of the labeled oxygen is found on the carbon atom bearing the two hydrogen atoms, although some product is formed in which the labeled oxygen is bonded to the carbon atom bearing the methyl group(s). These results are consistent with a mechanism in which hydroxide ion attacks the epoxide preferentially but not

Solutions Manual for Perspectives on Structure and Mechanism in Organic Chemistry, Third Edition.
Felix A. Carroll.

exclusively at the less highly substituted carbon atom by an S_N2 pathway.

H_2O^* slow ⟶ ⇌ H_2O^*

$O^* = {}^{18}O$

6.4. $800\,cm^{-1}$ corresponds to $2.5\times10^{13}\,s^{-1}$ (cf. Gordon, A. J.; Ford, R. A. *The Chemist's Companion*; John Wiley & Sons: New York), so 10^{13} is a ballpark estimate for the frequency of a molecular vibration. A rate constant that corresponds to this frequency is $1/(2.5\times10^{13}) = 4\times10^{12}\,s^{-1}$. If $\Delta G^\ddagger = RT$, then

$$k_r = \frac{\kappa \mathbf{k} T}{h} e^{-1} = 2.3\times10^{12}\ s^{-1}$$

6.5. Twenty-four hours is 86,400 s. The half-life, $t_{1/2}$, equals $\ln(2)/k$, so a half-life of 86,400 s equals a rate constant of $8\times10^{-6}\,s^{-1}$.[1] Here, τ is 124,675.3 s. Taking the natural logarithm of both sides of equation 6.44 and rearranging terms give

$$\Delta G^\ddagger = RT \ln\left(\frac{\mathbf{k} T \tau}{h}\right)$$

which equals 24.3 kcal/mol at 24 °C. Therefore, a first-order reaction with a half-life of 24 hr at 24 °C has a $\Delta G^\ddagger$ of about 24 kcal/mol.

6.6. Plot $-R\ln2/(1/T_2 - 1/T_1)$, where $T_2 = T_1 + 10\,°C$, to determine the E_a values for which the generalization is correct. At room temperature, E_a is ca. 12.5 kcal/mol. The E_a required for the generalization to be valid increases with the initial temperature.

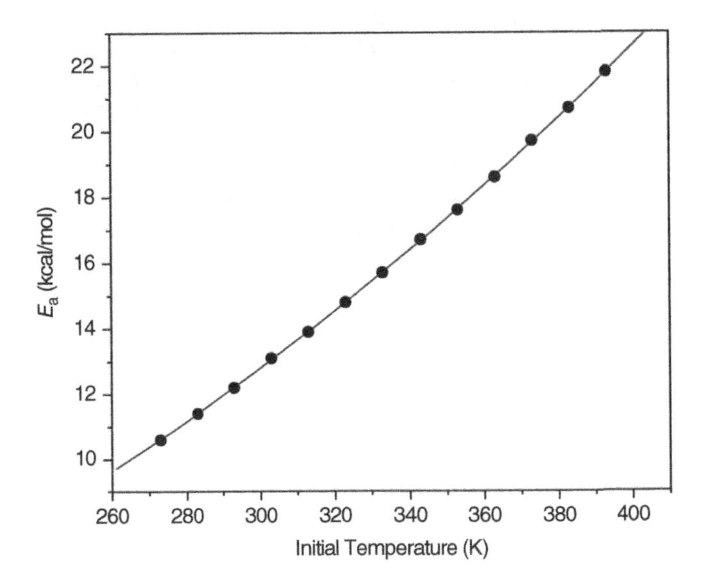

[1] The inverse of a first-order rate constant is called the *lifetime*, τ (see Chapter 12).

6.7. Bartlett, P. D.; Trachtenberg, E. N. *J. Am. Chem. Soc.* **1958**, *80*, 5808. For compound A, $\Delta H^{\ddagger}$ is 10.2 kcal/mol and $\Delta S^{\ddagger}$ is −30 eu. For compound B, the respective values are 31.3 kcal/mol and +25.4 eu. Thus, a lower temperature favors the reactant with the smaller activation enthalpy, while the higher temperature favors the reactant for which the activation entropy is more positive.

6.8. Frost, A. A.; Pearson, R. G. *Kinetics and Mechanism*, 2nd ed.; John Wiley & Sons: New York, 1961; pp. 101 and 105. Chung, Y.-S.; Duerr, B. F.; Nanjappan, P.; et al. *J. Org. Chem.* **1988**, *53*, 1334. Apparently the entropy of the transition structure is very nearly the same as that of the dimer. Therefore, $\Delta S^{\ddagger}$ for fragmentation is near zero. The entropy of the dimer is much less than that of two dicyclopentadiene molecules, however, so $\Delta S^{\ddagger}$ for dimerization is quite negative.

6.9. ΔG° is −10.9 kJ/mol. Nakazawa, J.; Sakae, Y.; Aida, M.; et al. *J. Org. Chem.* **2007**, *72*, 9448.

6.10. Hawkinson, D. C.; Wang, Y. *J. Org. Chem.* **2007**, *72*, 3592. $\delta\Delta G^{\ddagger} = RT\ln(12) = 1.47$ kcal/mol

$$\ln\frac{k_C}{k_B} = \ln\frac{[C]}{[B]} = \frac{\delta\Delta G^{\ddagger}}{RT}, \text{ so } \delta\Delta G^{\ddagger} = RT\ln(12) = 1.47 \text{ kcal/mol.}$$

6.11. Harada, N.; Saito, A.; Koumura, N.; Roe, D. C.; Jager, W. F.; Zijlstra, R. W. J.; de Lange, B.; Feringa, B. L. *J. Am. Chem. Soc.* **1997**, *119*, 7249 reported E_a and $\Delta H^{\ddagger}$ values of 21.5 ± 0.3 and 20.8 ± 0.3 kcal/mol, respectively.

6.12. LeFevre, G. N.; Crawford, R. J. *J. Org. Chem.* **1986**, *51*, 747. Both reactions are proposed to occur through bond homolysis leading to a diradical intermediate. The vinyl function in **101** stabilizes the diradical intermediate and lowers the activation enthalpy relative to cyclopropane. C–C bond stretching in the transition state leads to a positive activation entropy.

101

103

Conjugation with the π system requires loss of vinyl group rotational freedom

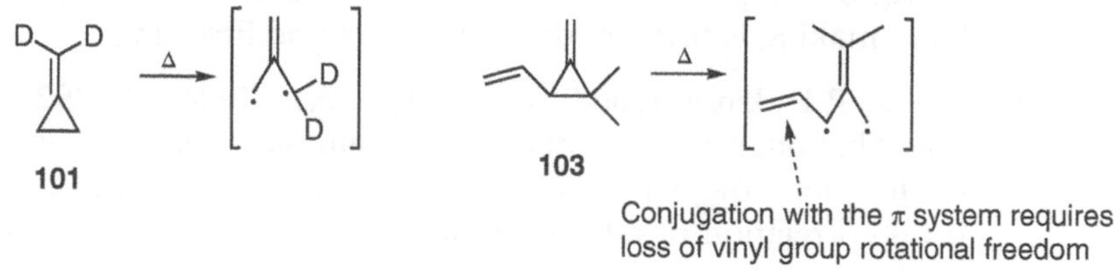

The additional vinyl group in **103** provides further stabilization of the diradical intermediate and lowers the activation energy even more. However, that stabilization requires the second vinyl group to become coplanar with the diradical π system. Loss of vinyl group rotational freedom in the transition state produces a negative activation entropy.

6.13. Kluger, R.; Brandl, M. *J. Org. Chem.* **1986**, *51*, 3964. Plot log k/T versus $1/T$. The slope is −4871.6, so $\Delta H^{\ddagger} = -4871.6 \times -4.576/1000 =$ 22.3 kcal/mol. $\Delta S^{\ddagger} = R \times 2.303 \times \log(k/T) + \Delta H^{\ddagger}/T - R \times 2.303 \times \log(\kappa\mathbf{k}/h) = -4.05$ eu at 25°. Alternatively, plot ln k versus $1/T$ to calculate E_a, which is found to be 22.9 kcal/mol. Then $\Delta H^{\ddagger}$ is estimated from the formula $\Delta H^{\ddagger} = E_a - RT = 22.3$ kcal/mol at 25°, and $\Delta S^{\ddagger} = 4.576 \log A - 60.53 = -3.9$ eu. Since $\Delta S^{\ddagger}$ is negative, the mechanism may be concerted. (See Chapter 9.) Such a mechanism is shown below:

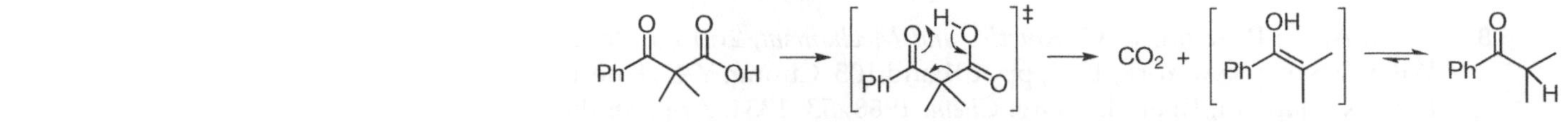

$$\text{Ph} \rightarrow [\ ]^{\ddagger} \rightarrow CO_2 + [\ ] \rightleftharpoons \text{Ph}$$

6.14. Bartlett, P. D.; Wu, C. *J. Org. Chem.* **1985**, *50*, 4087. The data suggest that the entropy is decreased in the transition structure, so a nondissociative mechanism is favored.

6.15. Jones, J. M.; Bender, M. L. *J. Am. Chem. Soc.* **1960**, *82*, 6322. This is a ß secondary effect. For the dissociation, $K_H/K_D = 1.44$.

6.16. Pocker, Y. *Proc. Chem. Soc.* **1960,** 17. This should be an inverse 2° isotope effect, so k_H/k_D should be less than 1. The literature value is 0.98.

6.17. Choe, J.-I.; Srinivasan, M.; Kuczkowski, R. L. *J. Am. Chem. Soc.* **1983**, *105*, 4703. There is an inverse 2° isotope effect, consistent with change from sp^2 to sp^3 hybridization at the transition state. Since the two isotope effects are identical, the data suggest nearly the same extent of bonding to both of the carbon atoms in the transition structure.

6.18. In ethanol solution, stabilization of the carboxylate ion by solvent is less effective than is the case in water. Therefore, the effect of a substituent on the benzene ring is more significant in ethanol than in water.

6.19. $\log \frac{k}{k_0} = \log k - \log k_0 = \sigma\rho$, so $\log k = \sigma\rho + c$, where c is a constant.

6.20. Overman, L. E.; Petty, S. T. *J. Org. Chem.* **1975**, *40*, 2779. The value calculated for ρ is 1.76. The investigators concluded that negative charge develops on both sulfur atoms in the transition state because bond making is further advanced than bond breaking.

6.21. Dietze, P. E.; Underwood, G. R. *J. Org. Chem.* **1984**, *49*, 2492.
a. and b. Both rate constants and pK_a values correlate better with σ^-. (Note: plot $-pK_a$ values.) The investigators report that there is a good correlation of rate constants with σ^-, giving $\rho^- = -2.54$, suggesting that the phenoxide undergoes significant change in charge distribution. Therefore, the N–Cl bond is thought to be substantially broken in the transition state, consistent with nucleophilic attack on chlorine.

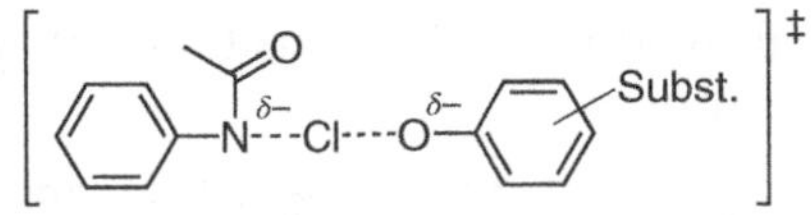

$$[\ \text{N}^{\delta-}\cdots\text{Cl}\cdots\text{O}^{\delta-}\ \text{Subst.}\ ]^{\ddagger}$$

The value of ρ for ionization was found to be 2.8 based on the pK_a values measured under the reaction conditions.

6.22. Whitworth, A. J.; Ayoub, R.; Rousseau, Y.; et al. *J. Am. Chem. Soc.* **1969**, *91*, 7128. At any one temperature, $\log(k/k_0) = (-0.91 \pm 0.03) \times \sigma$, so the addition is strongly electrophilic. That is, the reaction is facilitated by electron-donating substituents on the carbon–carbon double bond.

6.23. Hill, R. K.; Conley, R. T.; Chortyk, O. T. *J. Am. Chem. Soc.* **1965**, *87*, 5646.

$(CH_3)_3C$–C(=N–OH)–CH_3 $\xrightarrow{\text{PPA}}$ $(CH_3)_3CNH$–C(=O)–CH_3

CH_3C(Ph)(CH_3)–C(=N–OH)–Ph $\xrightarrow{\text{PPA}}$ Ph(CH_3)$_2$C–NH–C(=O)–Ph

$(CH_3)_3C$–C(=N–OH)–CH_3 + CH_3C(Ph)(CH_3)–C(=N–OH)–Ph $\xrightarrow{\text{PPA}}$ $(CH_3)_3CNH$–C(=O)–CH_3 + Ph(CH_3)$_2$C–NH–C(=O)–Ph + Ph(CH_3)$_2$C–NH–C(=O)–CH_3 + $(CH_3)_3CNH$–C(=O)–Ph

a. The observation of cross products strongly suggests that at least some product formation occurs through a dissociative pathway.
b. Optically active oximes that could produce achiral carbocations upon dissociation were investigated in the same study. Observation of racemic Beckmann rearrangement product is consistent with a dissociative pathway.
c. The dissociative pathway is most likely for those oximes with structural features that facilitate carbocation formation.

6.24. The literature value based on additional data is 0.94, but the data cited in the problem give a value of 1.09. Maclin, K. M.; Richey, H. G., Jr. *J. Org. Chem.* **2002**, *67*, 4370.

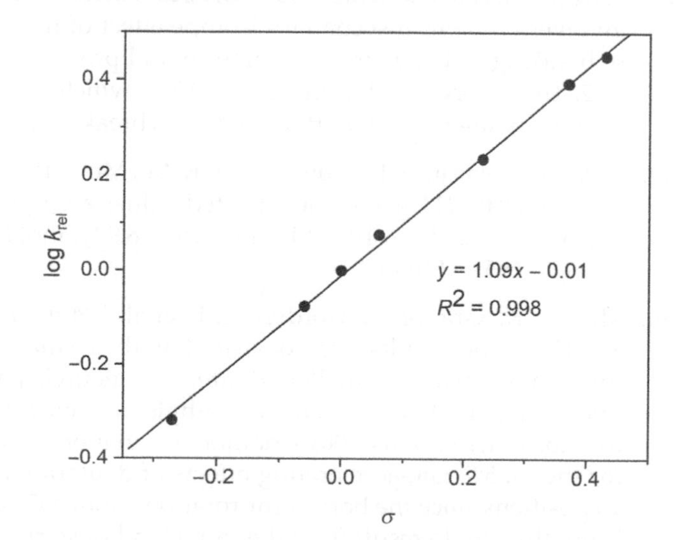

6.25. For each temperature, the k for a substituent is a constant multiple of the k for S = H. Therefore, an Eyring or Arrhenius plot will have the same slope for each substituent, so all will have the same activation energies and $\Delta H^{\ddagger}$ values. Therefore, the differences in reactivity for the compounds must arise from differences in $\Delta S^{\ddagger}$. Kim, S. S.; Choi, W. J.; Zhu, Y.; et al. *J. Org. Chem.* **1998**, *63*, 1185.

6.26. Plot $-\log K_i$ versus σ_p. A good correlation is obtained, with $-\log K_i = 2.87\sigma_p - 1.47$. Using 0.54 as the σ_p value of CF_3 gives $-\log K_i = 0.08$, making $K_i = 0.83$ nM. The literature value is $K_i = 0.8$ nM. Romero, F. A.; Hwang, I.; Boger, D. L. *J. Am. Chem. Soc.* **2006**, *128*, 14004.

6.27. Reddy, S. R.; Manikyamba, P. *J. Chem. Sci.* **2006**, *118*, 257. Determine the value of ρ for each temperature, as shown in the figure below. Then plot ρ versus the temperature to obtain the correlation $\rho = -0.0372T + 11.725$. The value of ρ is 0 when $T = 315.2$ K.

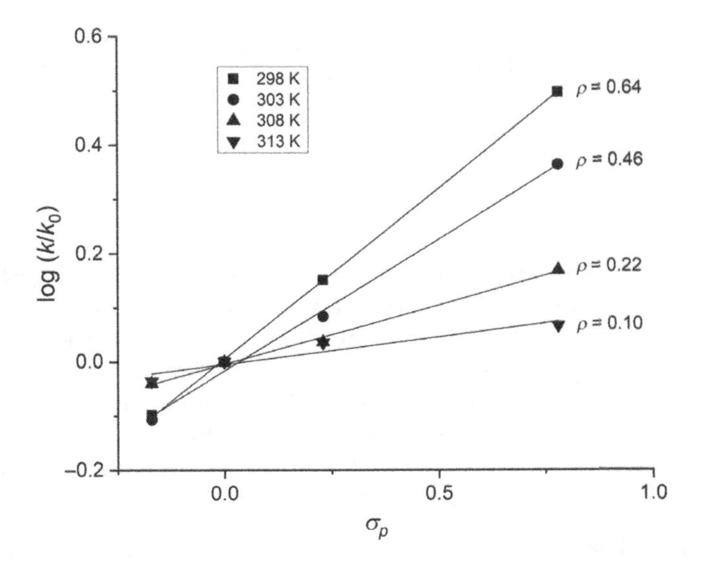

6.28. Seltzer, S. *J. Am. Chem. Soc.* **1961**, *83*, 2625. Each C–N bond-breaking involves only one α-secondary isotope effect of the sp^3 to sp^2 type, so breaking just one bond at a time should produce a KIE of about 1.12. The observed KIE is inverse of 0.787, which is 1.27. Therefore, the results suggest that both C–N bonds break at the same time.

6.29. Data from Marquez, E.; Domínguez, R. M.; Mora, R. R.; et al. *J. Phys. Chem. A* **2010**, *114*, 4203. The reported values are $E_a = 196.3$ kJ/mol, $\log A = 14.0$, $\Delta H^{\ddagger} = -191.3$ kJ/mol, $\Delta S^{\ddagger} = 8.93$ J/mol·K, and $\Delta G^{\ddagger}$ at 330 °C = 185.9 kJ/mol.

6.30. Mislow, K.; Graeve, R.; Gordon, A. J.; et al. *J. Am. Chem. Soc.* **1963**, *85*, 1199. The results are consistent with a smaller barrier for rotation of a pair of "smaller" deuterated methyls past each other than for a pair of nondeuterated methyls (relative rate constant for rotation of 1.13 versus 1.00). Another explanation must be operative for the ca. 5% rate-accelerating effects of deuteration in the 9 and 10 positions since the barrier for rotation about a C–C single bond is not thought to result from the same kind of steric effects.

6.31. King, F. J.; Searle, A. D.; Urquhart, M. W. *Org. Process Res. Dev.* **2020**, *24*, 2915.

The investigators carried out a crossover experiment using unlabeled ranitidine and ranitidine labeled with *both* deuterium and ^{15}N as shown on the left below. The reaction produced only unlabeled NDMA and doubly labeled NDMA (shown on the right below), suggesting that the *N*-nitroso compound was formed exclusively through an intramolecular reaction.

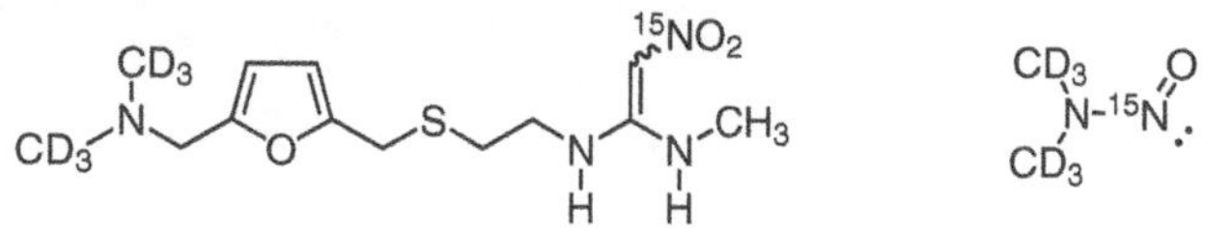

6.32. Choi, S.; Breugst, M.; Houk, K. N.; et al. *J. Org. Chem.* **2014**, *79*, 3572. The delta isotope effect was explained in terms of hyperconjugation from the CH_3 or CD_3 group of CX_3 in the carbocation.

CHAPTER 7

Acid and Base Catalysis of Organic Reactions

7.1. Barlin, G. B.; Perrin, D. D. *Quart. Rev. Chem. Soc.* **1966**, *20*, 75.

$$pK_a = 4.20 - \Sigma\sigma = 4.20 - (-0.37 + 0.115) = 4.46$$

7.2. Taft, R. W.; Bordwell, F. G. *Acc. Chem. Res.* **1988**, *21*, 463.

$$\Delta G^\circ = -RT \ln K_a = 2.303RT(-\log K_a) = 2.303RT\,pK_a$$

Therefore, $pK_a = \Delta G^\circ/1.364$ (where ΔG° values are in kcal/mol). Using the data from Table 7.4, the pK_a values of acetic and propionic acid in the gas phase are found to be 250.4 and 249.6 kcal/mol, respectively. This is a difference of 0.8 kcal/mol or 0.59 pK units, with propionic acid being more acidic. The data in Table 7.1 indicate that the pK_a values of acetic and propionic acid in aqueous solution are 4.76 and 4.87, respectively. This is a difference of 0.11 pK units, with acetic acid being more acidic. The magnitude of the pK_a values is smaller in solution because the solvent helps stabilize the ion (resulting in much smaller pK_a values), so the effect of the substituent is less significant. Furthermore, propionic acid is more acidic in the gas phase because the replacement of a ß-hydrogen with a methyl group provides greater dispersal of the negative charge in the gas phase. In solution, however, the methyl substituent causes greater disruption of the solvent shell around the carboxylate ion, reducing the acidity of the larger acid.

Solutions Manual for Perspectives on Structure and Mechanism in Organic Chemistry, Third Edition.
Felix A. Carroll.

7.3. The first substitution (from methanol to ethanol) reduces the calculated ΔG° value by 3.01 kcal/mol; the second (from ethanol to isopropyl alcohol) by 1.97 kcal/mol, and the third (from isopropyl alcohol to *t*-butyl alcohol) by 1.23 kcal/mol. Each successive substitution has less effect than the previous one because the effect of a methyl group in delocalizing the charge is less significant in a large alkyl group (where there is already significant charge delocalization) than in a smaller group where the charge is not so extensively delocalized.

7.4. Wheeler, O. H. *J. Am. Chem. Soc.* **1957**, *79*, 4191.

For cyclobutanone, $K_d = 1.11$. Angle strain is relieved by conversion of the sp^2 (120° bond angle preferred) carbonyl carbon to sp^3 (109.5° bond angle preferred) hybridization.

For cyclopentanone, $K_d = 15.1$. Angle strain is not relieved appreciably by hemiacetal formation, and some steric hindrance between the methoxy group and the ring hydrogen atoms results from hemiacetal formation.

For cyclohexanone, $K_d = 2.1$. Some angle strain is reduced by hemiacetal formation, but some intramolecular van der Waals repulsion is introduced.

7.5. Use equation 7.15 and substitute $-\mathrm{p}K_b = \log K_b = \mathrm{p}K_{BH+} - 14$.

7.6. Wiseman, J. S.; Abeles, R. H. *Biochem.* **1979**, *18*, 427.

a. Hydrolysis of the amide function can occur by mechanisms analogous to those shown in Figures 7.28 and 7.30.

b. Cyclopropanone exists almost exclusively in the form of the hydrate because of the relief of angle strain upon conversion of the sp^2- to sp^3-hybridized carbonyl carbon atom in the cyclopropanone ring.

7.7. Cardwell, H. M. E.; Kilner, A. E. H. *J. Chem. Soc.* **1951**, 2430 and references therein. In each case, reaction occurs on the more substituted α carbon atom.

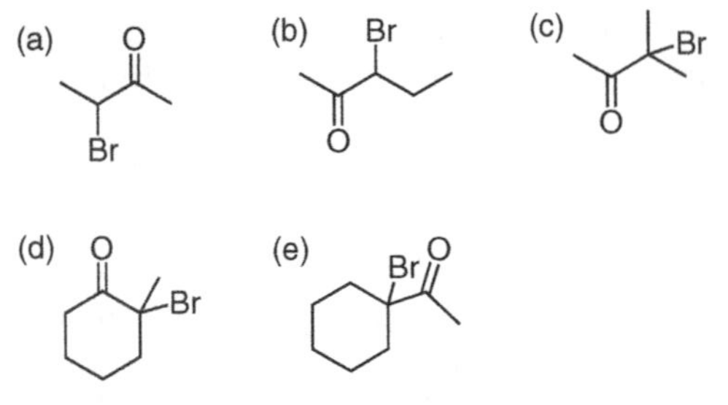

7.8. Levine, R.; Hauser, C. R. *J. Am. Chem. Soc.* **1944**, *66*, 1768. Base-promoted enolization occurs on the methyl group, the less-substituted α carbon atom.

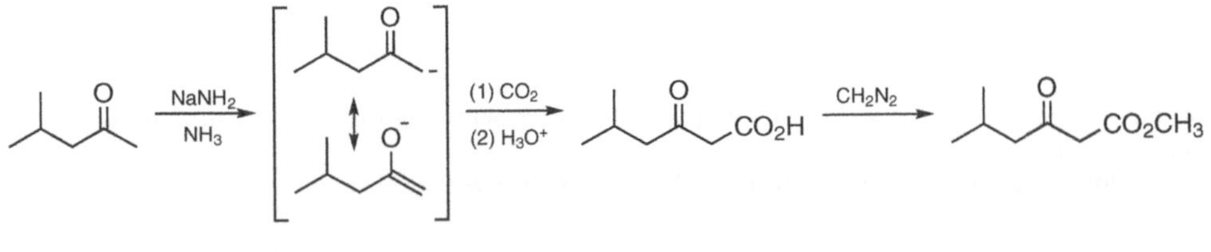

7.9. Gutsche, C. D.; Redmore, D.; Buriks, R. S.; et al. *J. Am. Chem. Soc.* **1967**, *89*, 1235. The reaction is subject to general base catalysis, as indicated by the linear correlation of log k with pK_{BH^+}. However, there is also a steric component to the catalysis because there are separate correlations for pyridine bases with 0, 1, and 2 (one compound only) methyl groups ortho to the pyridine nitrogen.

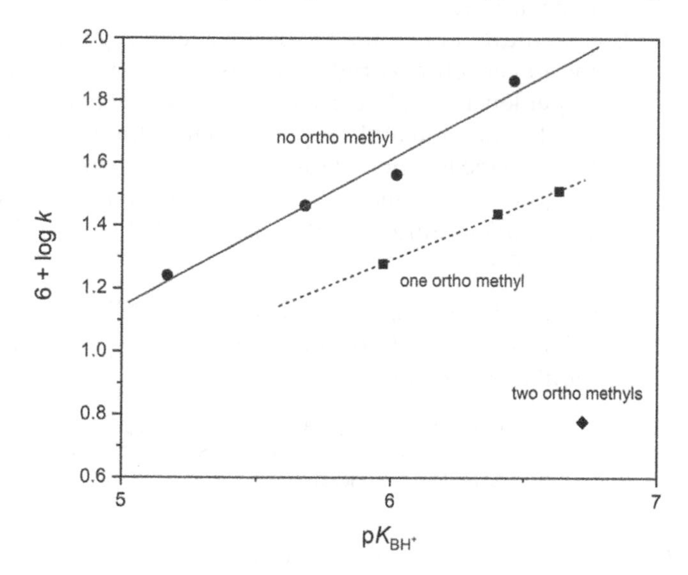

7.10. Smith, W. T.; McLeod, G. L. *Org. Syn. Coll. Vol. IV* **1963**, 345. Also see Walker, J.; Wood, J. K. *J. Chem. Soc.* **1906**, *89*, 598. There are multiple haloform reactions.

NaOCl / aq. NaOH → (Cl, Cl) → aq. NaOH → Cl_2HC– … –O^- → NaOCl / aq. NaOH → H^+ → HO– … –OH

7.11. Stefanidis, D.; Cho, S.; Dhe-Paganon, S.; et al. *J. Am. Chem. Soc.* **1993**, *115*, 1650. The formate ester should form the more stable tetrahedral intermediate for reasons similar to those for the more stable addition product for hydration of an aldehyde than a ketone.

7.12. Sørensen, P. E.; Jencks, W. P. *J. Am. Chem. Soc.* **1987**, *109*, 4675.

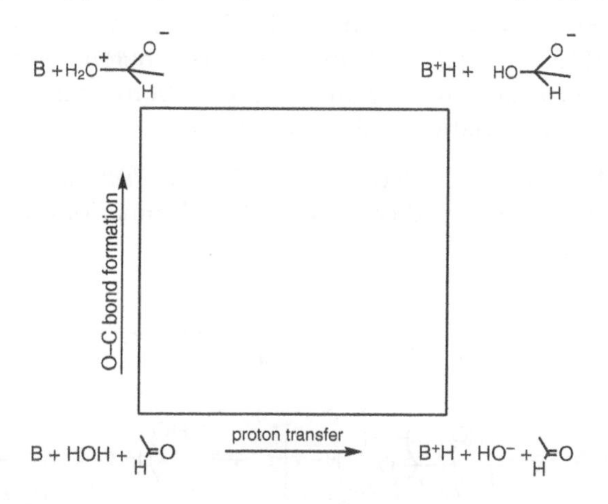

a. A stronger base raises the energy of the left-hand side of the diagram. As a result, the location of the transition structure on the diagram moves toward the bottom of the figure. This movement is the result of movement toward the lower right corner (that is perpendicular to the reaction coordinate) and toward the lower left corner (a Hammond effect) resulting from a more exothermic reaction.

b. Substitution of better electron-donating groups would stabilize the lower left and right corners of the diagram relative to the upper left and right corners. The net effect of the resulting movement parallel to and perpendicular to the reaction coordinate is to move the location of the transition structure to the right on the diagram. For a more detailed discussion of these effects, see page 4687 of Sørensen, P. E.; Jencks, W. P. *J. Am. Chem. Soc.* **1987**, *109*, 4675.

7.13. Bell, R. P.; Baughan, E. C. *J. Chem. Soc.* **1937**, 1947. Intermolecular reaction converts two molecules of dihydroxyacetone into a dihemiketal structure.

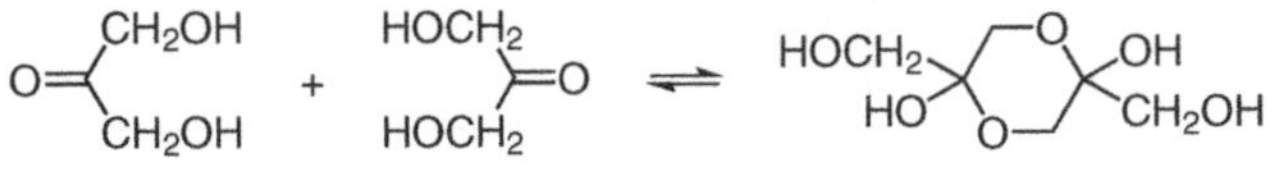

7.14. Hurd, C. D.; Saunders, W. H., Jr. *J. Am. Chem. Soc.* **1952**, *74*, 5324. Intramolecular hemiacetal formation reduces the concentration of carbonyl groups in solution:

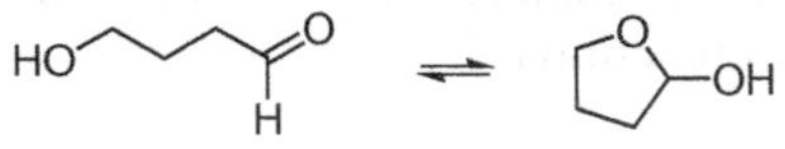

For hydroxyaldehydes having the formula $HO(CH_2)_nCHO$, the fraction of free aldehyde groups varies with the chain size because 5- and 6-membered rings are formed more readily than are other size rings. The fraction of free aldehyde varies with n as follows: 3, 0.114; 4, 0.061; 5, 0.85; 7, 0.80; 8, 0.91.

7.15. Drumheller, J. D.; Andrews, L. J. *J. Am. Chem. Soc.* **1955**, *77*, 3290. The alcohol should be recovered with the same configuration as in the acetal since the bond from oxygen to the chiral center is not broken in the hydrolysis reaction.

7.16. See the discussion in Bender, M. L.; Chen, M. C. *J. Am. Chem. Soc.* **1963**, *85*, 30; *ibid.*, 37 and references therein. Steric hindrance makes attack of a nucleophile on the carbonyl carbon atom slower than the dissociation of the acylium ion.

7.17. Reimann, J. E.; Jencks, W. P. *J. Am. Chem. Soc.* **1966**, *88*, 3973. Note the general acid catalysis in the elimination of water in the second step.

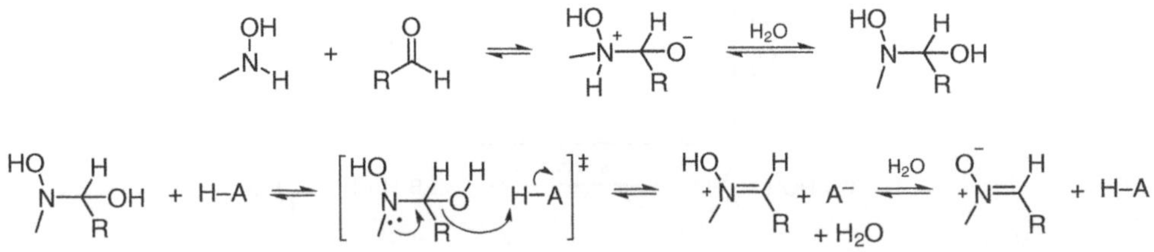

7.18. Olson, A. R.; Youle, P. V. *J. Am. Chem. Soc.* **1951**, *73*, 2468. Carbonate ion produces general base catalysis for a pathway involving breaking of the acyl carbon—ß-oxygen bond. Acetate ion produces nucleophilic catalysis of a pathway involving breaking the bond between the ß-carbon atom and the oxygen atom.

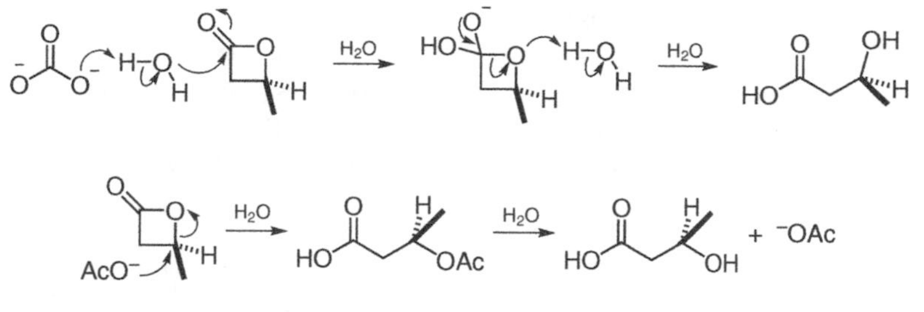

7.19. Hoz, S.; Livneh, M.; Cohen, D. *J. Org. Chem.* **1986**, *51*, 4537. Plot log k versus log K (that is, $-\mathrm{p}K_a$). A good linear correlation is observed, with $\alpha = 0.96$.

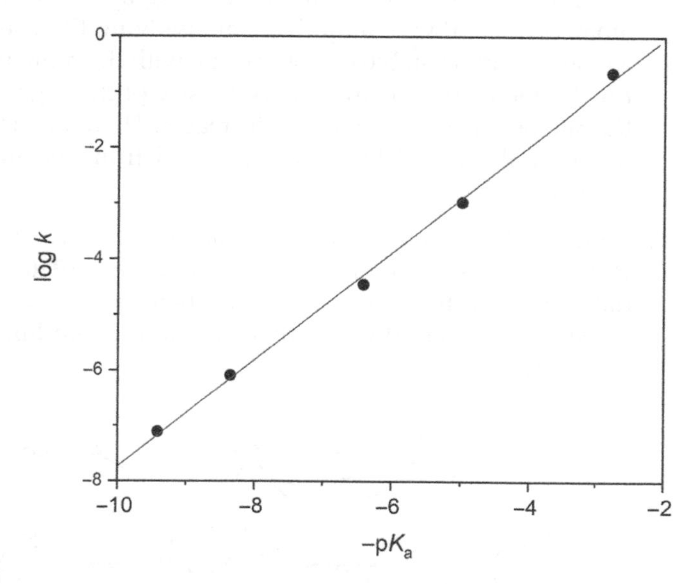

7.20. Zaugg, H. E.; Papendick, V.; Michaels, R. J. *J. Am. Chem. Soc.* **1964**, *86*, 1399; see also Johnson, S. L. *Adv. Phys. Org. Chem.* **1967**, *5*, 237. The observation of **19** supports the existence of a tetrahedral intermediate in the reaction. Both products are formed by intramolecular S_N2 reactions. The major product is formed by an S_N2 reaction of the phenol attacking the alkyl bromide moiety after conversion of the lactone to the amide. The minor product is formed by attack of the alkoxide moiety of the tetrahedral intermediate of the acyl substitution reaction with the alkyl bromide group.

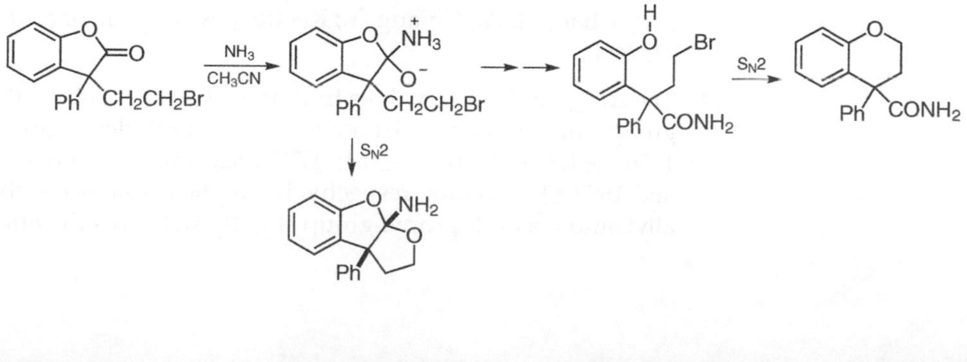

7.21. Hauser, C. R.; Adams, J. T. *J. Am. Chem. Soc.* **1944**, *66*, 345.

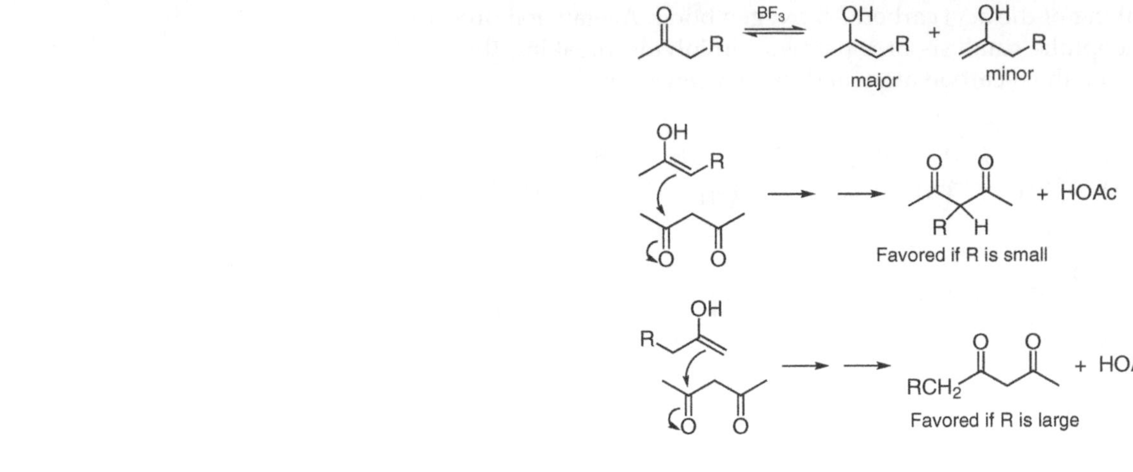

The acylation occurs on the more highly substituted α carbon atom, consistent with a mechanism involving BF_3-catalyzed enolization. The more stable enol is the one with the more highly substituted double bond. However, as the size of the R group increases, the rate of reaction of this enol decreases. Therefore, reaction of the minor enol, formed by proton removal from the methyl group, becomes competitive.

7.22. Long, F. A.; Pritchard, J. G. *J. Am. Chem. Soc.* **1956**, *78*, 2663; Pritchard, J. G.; Long, F. A. *J. Am. Chem. Soc.* **1956**, *78*, 2667. The data are consistent with an A-1 mechanism in which protonated epoxide is converted to a carbocation in the rate-limiting step for the reaction.

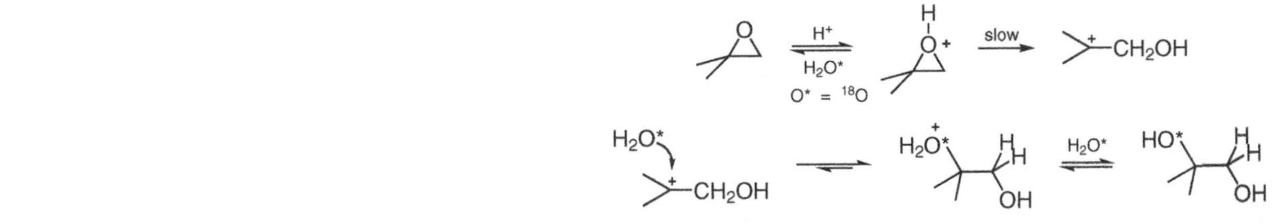

Because 1° carbocations are so much less stable than 2° or 3° carbocations, it seems likely that the minor products in the acid-catalyzed hydrolyses do not arise from opening of the protonated epoxide to 1° carbocations. Instead, attack of water on protonated epoxide may occur in competition with opening of epoxide to carbocation.

7.23. The increased acidity is consistent with the greater *s* character of the orbitals used to make the exocyclic bonds of the four-membered ring. Chang, J. A.; Chiang, Y.; Keeffe, J. R.; et al. *J. Org. Chem.* **2006**, *71*, 4460.

7.24. The ΔH_{acid} and ΔG_{acid} values from the *NIST Webbook* for the methyl group of propene, 1-butene, and methylcyclopropane are 1636.4 ± 1.3 and 1605.8 ± 0.4; 1724 ± 8.8 and 1690 ± 8.8; 1718 ± 8.4 and 1682 ± 8.8 kJ/mol, respectively. Allylic resonance stabilizes the allyl anion; a cyclopropyl group slightly stabilizes the anion.

7.25. Partial mechanisms are shown here. For a more details, see the reference cited below.

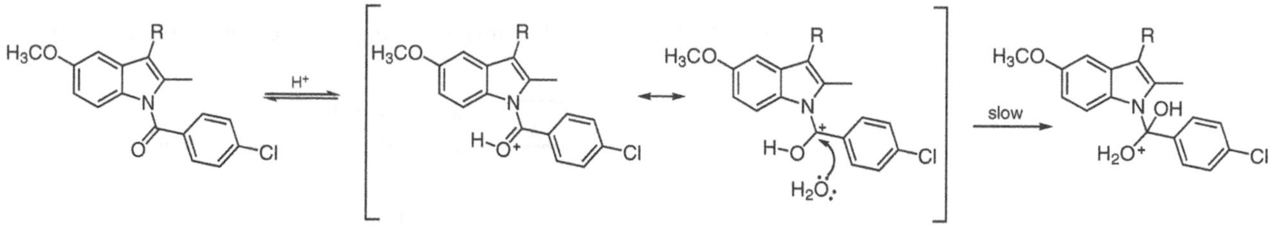

A-2 Mechanism for the Hydrolysis of the Amide Group

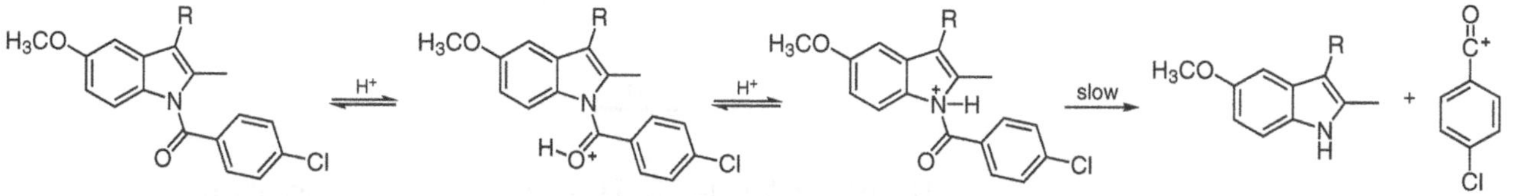

A-1 Mechanism for the Hydrolysis of the Amide Group

Adapted with permission from García, B.; Hoyuelos, F. J.; Ibeas, S.; et al. *J. Org. Chem.* **2006**, *71*, 3718. © 2006 American Chemical Society.

7.26. Plot GB values vs. σ^+. The value obtained for styrene is 195.2 kcal/mol. Harrison, A. G.; Houriet, R.; Tidwell, T. T. *J. Org. Chem.* **1984**, *49*, 1302.

7.27. There is a good correlation with $-H_0$ but not with $[H^+]$. This suggests that the transition state behaves as a protonated dimethoxymethane. McIntyre, D.; Long, F. A. *J. Am. Chem. Soc.* **1954**, *76*, 3240.

7.28. Intramolecular hydrogen bonding can stabilize the carboxylate group in the cis isomers. For the cis 1,3 isomer, hydrogen bonding can occur in a chair conformation, but for the cis 1,4 isomer a twist boat conformation is required. Chen, X.; Walthall, D. A.; Brauman, J. I. *J. Am. Chem. Soc.* **2004**, *126*, 12614.

7.29. Gattin, Z; Kovačević, B.; Maksić, Z. B. *Eur. J. Org. Chem.* **2005**, 3206. Not only does the compound have six amine functions, but protonation of the imine nitrogen leads to a symmetric cation with an aromatic ring:

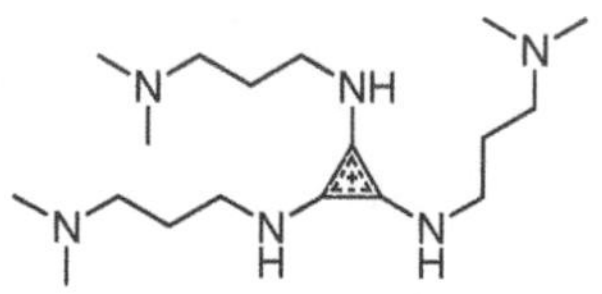

7.30. See the discussion in Gassman, P. G.; Zalar, F. V. *J. Am. Chem. Soc.* **1966**, *88*, 3070 and Bartlett, P. D.; Woods, G. F. *J. Am. Chem. Soc.* **1940**, *62*, 2933. The enolate is unstable because it would violate Bredt's rule.

7.31. Breslow, R. *Chem. Rec.* **2014**, *14*, 1174. Triphenylmethane (pK_a 30) is much more acidic than 1,2,3-triphenylcyclopropene (pK_a 50) because the 1,2,3-triphenylcyclopropenyl anion is antiaromatic.

7.32. Kütt, A.; Leito, I.; Kaljurand, I.; et al. *J. Org. Chem.* **2006**, *71*, 2829.

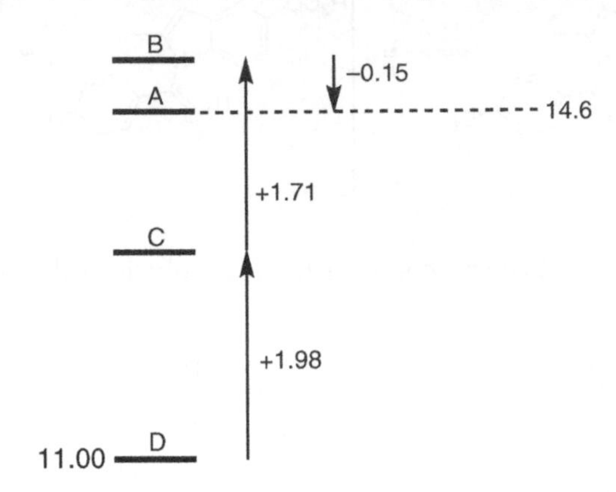

The pK_a of **A** is 14.57. (**D** is picric acid; **A** is saccharin.)

7.33. Gómez-Bombarelli, R.; Calle, E.; Casado, J. *J. Org. Chem.* **2013**, *78*, 6880 and 6868.

(a) $B_{AC}1$; (b) $A_{AC}1$; (c) $B_{Al}1$; (d) $A_{Al}1$; (e) $B_{Al}2$; (f) $A_{Al}2$; (g) $B_{AC}2$.

7.34. Douglas, J. E.; Campbell, G.; Wigfield, D. C. *Can. J. Chem.* **1993**, 71, 1841. The hydrolysis of **25** takes place entirely by the $B_{AC}2$ mechanism in spite of the steric hindrance of the α carbon of the ester. The hydrolysis of the even more hindered **26** takes place 5% of the time by the $B_{Al}2$ mechanism and 95% of the time by the $B_{AC}2$ mechanism.

7.35. Barclay, L. R. C.; Hall, N. D.; Cooke, G. A. *Can. J. Chem.* **1962**, 40, 1981. The reaction of **27** takes place entirely by the $B_{Al}2$ mechanism, so all of the label from water is in the methanol. The reaction of **28** takes place mostly by the $B_{Al}2$ mechanism but some occurs by the $B_{Ac}2$ pathway. Compound **29** reacts only by the $B_{Ac}2$ mechanism. These results indicate that both of the substituents at the 2 and 6 positions of the benzoate function must be *t*-butyl for there to be enough steric hindrance to stop the $B_{Ac}2$ pathway.

7.36. Bunton, C. A.; Konasiewicz, A. *J. Chem. Soc.* **1995**, 1354. Because of the stability of the triphenylmethyl cation, the reaction takes place nearly entirely by the $B_{Al}1$ mechanism under neutral conditions. As the concentration of base increases, some $B_{Ac}2$ reaction occurs.

CHAPTER 8

Substitution Reactions

8.1. Ingold, C. K. *Structure and Mechanism in Organic Chemistry*, 2nd ed.; Cornell University Press: Ithaca, 1969; pp. 457 ff.

a. Type III: large decrease. (The products of the reaction are CH_3I and $(CH_3)_3N$.)

b. Type IV: small decrease. (The products of the reaction are $CH_3N^+H_3$ and $(CH_3)_2S$.)

8.2.

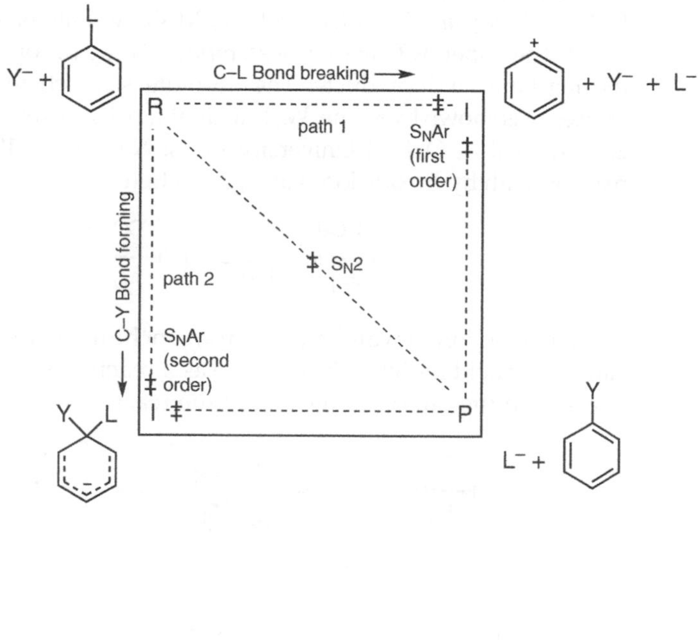

Solutions Manual for Perspectives on Structure and Mechanism in Organic Chemistry, Third Edition.
Felix A. Carroll.

8.3. Magid, R. M.; Welch, J. G. *J. Am. Chem. Soc.* **1968**, *90*, 5211. The results suggest that about 75% of the product is formed by the S_N2' pathway.

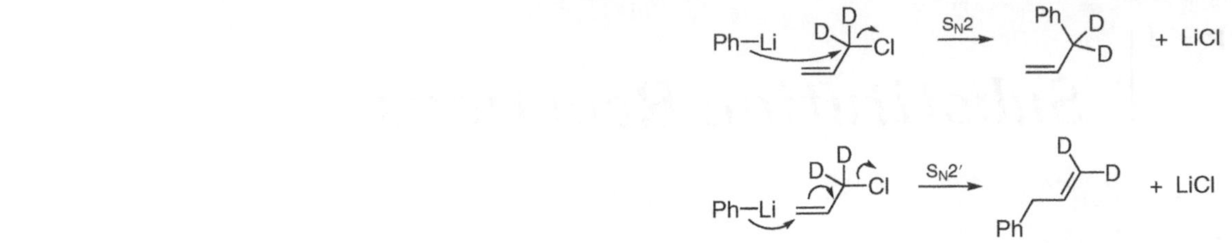

8.4. Smith, M. B.; Hrubiec, R. T.; Zezza, C. A. *J. Org. Chem.* **1985**, *50*, 4815. One product is formed by direct S_N2 reaction, while the other is formed by an S_N2' process. The larger the R group, the greater the steric barrier for the S_N2 pathway, so the greater is the yield of product formed by the S_N2' process. The S_N2' product is obtained in 8% yield when R is methyl, but 88% when R is *t*-butyl.

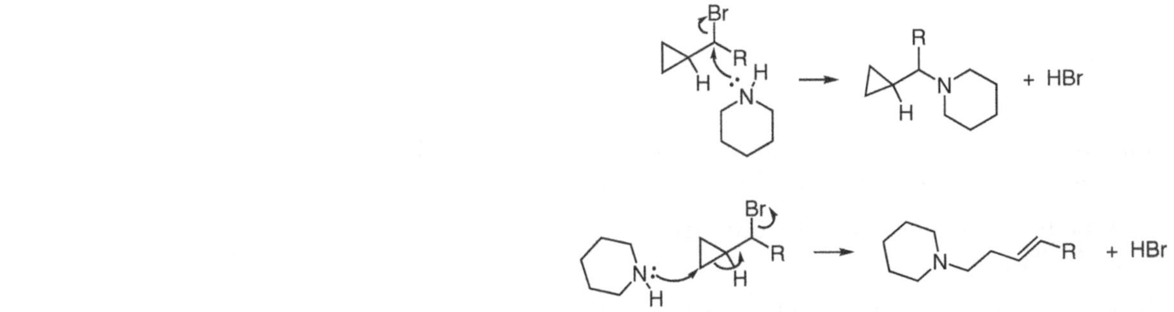

8.5. Cowdrey, W. A.; Hughes, E. D.; Ingold, C. K. *J. Chem. Soc.* **1937**, 1208; Winstein, S.; Lucas, H. J. *J. Am. Chem. Soc.* **1939**, *61*, 1576. See also Hine, J. *Physical Organic Chemistry*, 2nd ed.; McGraw-Hill: New York, **1962**; p. 143. The paper by Cowdrey et al. reported only the optical activity of the reactants and products. The absolute stereochemistry of these species was shown by Klyne, W.; Buckingham, J. *Atlas of Stereochemistry*, 2nd ed., Vol. 1; Oxford University Press: New York, **1978**, p. 5. The reaction with hydroxide ion is an S_N2 reaction.

The reaction in water results from anchimeric assistance by the carboxylic acid moiety (shown here as a reaction with the carboxylate ion) in the rate-limiting step of the reaction:

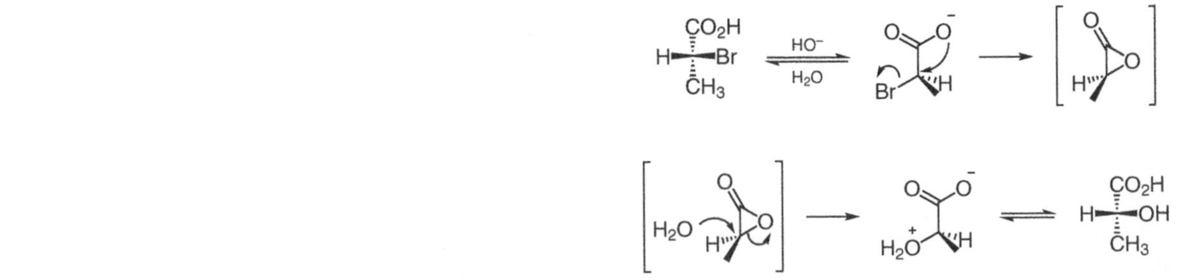

8.6. McKenzie, A.; Clough, G. W. *J. Chem. Soc.* **1910**, *97*, 2564. The reaction with $SOCl_2$ is an S_Ni reaction that proceeds with retention of configuration. The reaction with PCl_5 involves S_N2 reaction of

Cl^- with the initial product of reaction of the alcohol with PCl_5, so this reaction proceeds with inversion of configuration. For a discussion, see McRae, W. *Basic Organic Reactions*; Heyden & Son, Ltd.: London, 1973; pp. 115–116.

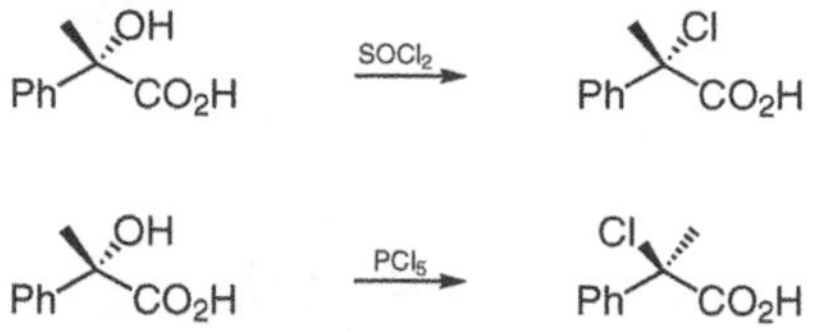

8.7. Brown, R. F.; van Gulick, N. M. *J. Org. Chem.* **1956**, *21*, 1046. The rate-limiting step is anchimeric assistance by the NH_2 group in the departure of the bromide ion. The larger the R groups, the greater the percentage of conformations of the reactant in which the amino and bromomethyl groups have the proper orientation for neighboring group participation.

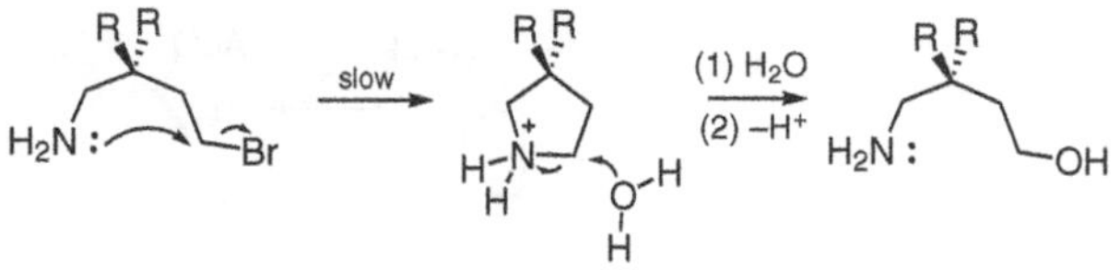

8.8. a. Cram, D. J. *J. Am. Chem. Soc.* **1949**, *71*, 3863. Also see Gould, E. S. *Mechanism and Structure in Organic Chemistry*; Holt, Rinehart and Winston: New York, **1959**; p. 576 and references therein. The stereochemical designations are evident in the Fischer projections for the reactant and products. For a review of the terminology, see Chapter 3.

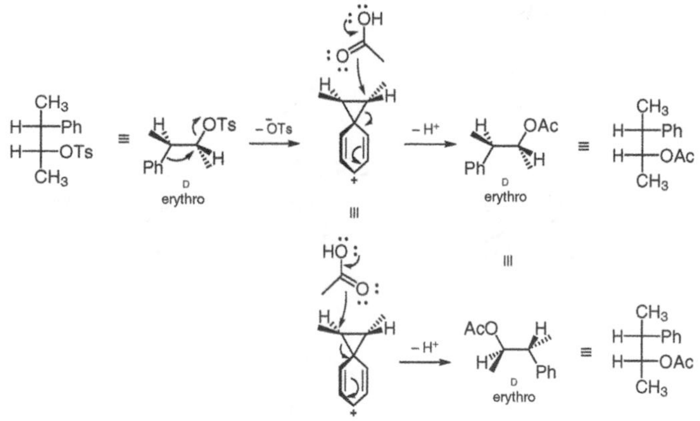

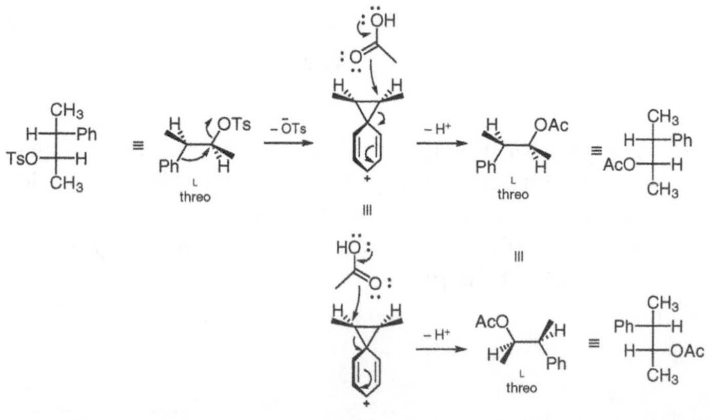

b. Cram, D. J.; Davis, R. *J. Am. Chem. Soc.* **1949**, *71*, 3875. The phenonium ion in this case does not have a plane of symmetry because one carbon atom bears a methyl group but the other carbon atom has an ethyl group. However, both carbon atoms are susceptible to nucleophilic attack by solvent.

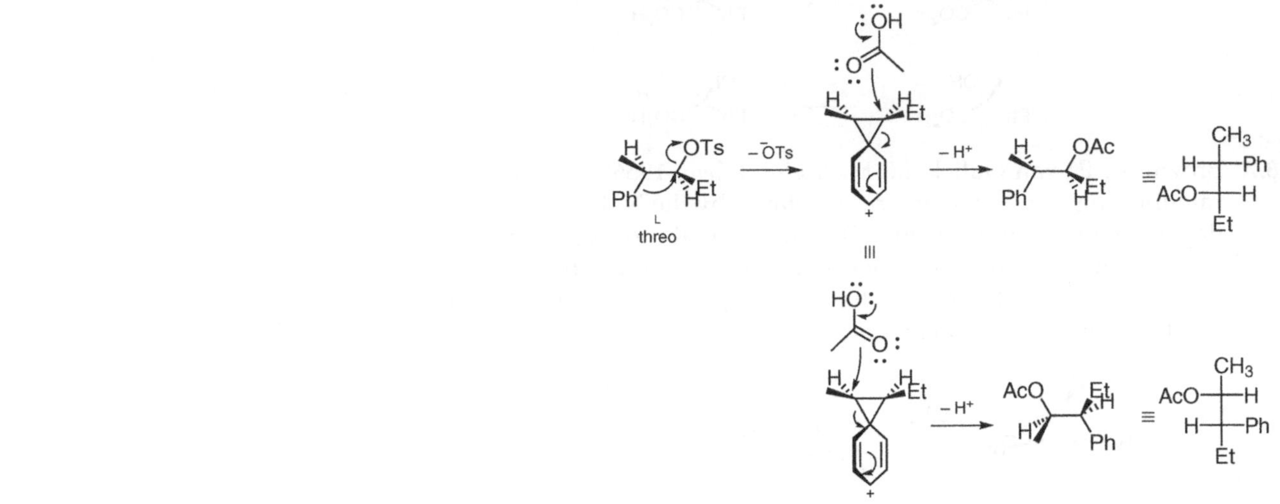

A product mixture that is the mirror image of that formed above would be obtained by reaction of a phenonium ion that is the mirror image of the phenonium ion shown above. Therefore, the reactant that would lead to that phenonium ion is the isomer of 3-phenyl-2-pentyl tosylate in which the methyl and ethyl groups reverse positions from those in L-*threo*-2-phenyl-3-pentyl tosylate.

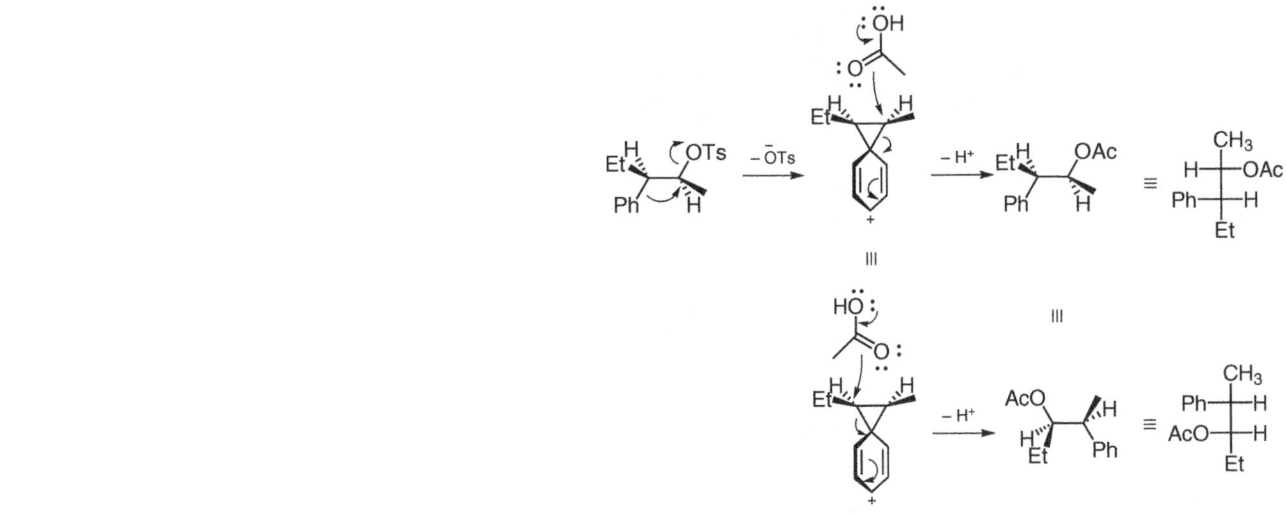

8.9. The plot for hexane should resemble that for the gas phase, but the energy should be slightly lower on the right side of the plot. Similarly, the plot for acetone solution should resemble that for aqueous solution but should be slightly higher in energy on the right side of the plot.

8.10. Winstein, S.; Lucas, H. J. *J. Am. Chem. Soc.* **1939**, *61*, 1581. The observation of (±)-*erythro*-3-bromo-2-butanol suggests that the ester function is cleaved by acid-catalyzed hydrolysis. This leads to the following mechanism (shown with Maehr notation):

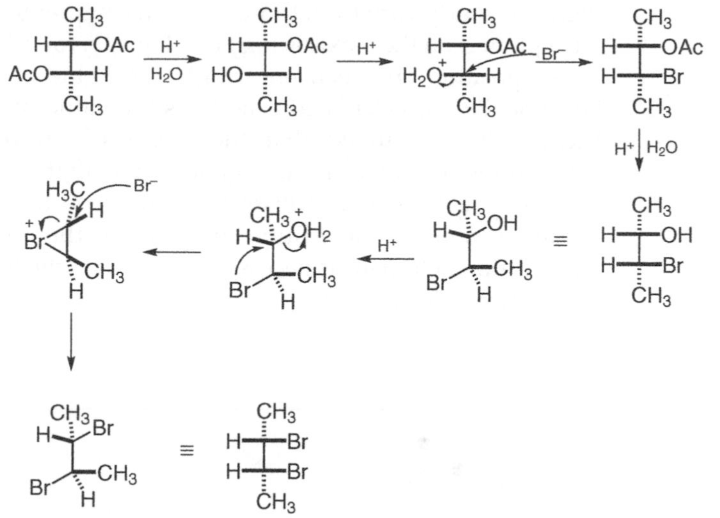

8.11. Chapman, J. W.; Strachan, A. N. *Chem. Commun.* **1974**, 293; see also Stock, L. M. *Prog. Phys. Org. Chem.* **1976**, 12.

$$k_{\text{toluene}}/k_{\text{benzene}} = (2\times41+2\times2.1+51)/6 = 23$$
(rounded to the nearest integer).

8.12. Masci, B. *J. Org. Chem.* **1985**, *50*, 4081.

a. Partial Rate Factors: Effect of [21C7] on Nitration of Anisole

	[21C7]			
	0.0 M	0.0082 M	0.032 M	0.131 M
f_o	5.20×10^3	5.01×10^3	4.23×10^3	3.97×10^3
f_p	2.92×10^3	1.96×10^4	5.62×10^4	1.57×10^5

b. The results suggest that O_2N^+ ions complexed with 21C7 act as electrophiles in the reaction and that complexed electrophiles are much more sensitive to substituent effects than are uncomplexed electrophiles. In addition, the relative magnitude of f_p/f_o increases dramatically with [21C7] because of the much greater steric effect of the complexed nitronium ion.

8.13. Brown, H. C.; Jensen, F. R. *J. Am. Chem. Soc.* **1958**, *80*, 2296.

a. Relative Rate Data for Benzoylation of Benzene and Derivatives in Benzoyl Chloride Solution at 25 °C.

Compound	*k* (relative)
Benzene	1
Toluene	110
t-Butylbenzene	72.4

b. $f_o = 30.7; f_m = 4.8; f_p = 589$
For example, $f_o = 0.03 \times 110 \times 9.3 = 30.7$

c. Plotting log k/k_0 versus σ^+ gives a good correlation with $\rho = -10$. The literature value is −9.57, but the values of σ^+ used in the literature study vary from those listed in Chapter 6. Note that the term plotted on the *y*-axis is log *f* (i.e. log k/k_0 for rates of reaction at *one* position on a benzene ring). The value of f_p for *t*-butyl benzene (393) is calculated exactly as the value of f_p for toluene because it is assumed that the yields of ortho- and meta-substitution products is negligible. Note that the figure also includes log *f* for meta substitution. Because the σ^+ and log f_p values of methyl and *t*-butyl are similar, the value for meta substitution of toluene is necessary to validate the linear relationship between log *f* and σ^+.

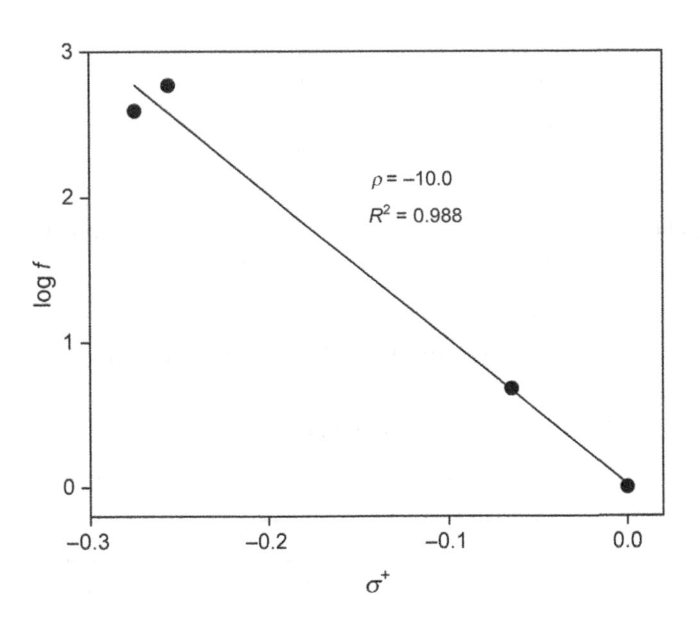

8.14. a. Himeshima, Y.; Kobayashi, H.; Sonoda, T. *J. Am. Chem. Soc.* **1985**, *107*, 5286. This is a first-order S_NAr reaction.

OSO_2CF_3 (CH3)3Si– –Si(CH3)3 —HOCH2CF3→ (CH3)3Si– + –Si(CH3)3 + $F_3CSO_3^-$ —HOCH2CF3→ OCH_2CF_3 (CH3)3Si– –Si(CH3)3

b. The reaction proceeds through formation of a phenyl carbocation that can capture solvent to give the first product or can undergo successive rearrangements involving an ortho *t*-butyl group to give another carbocation that captures solvent to give the second product.

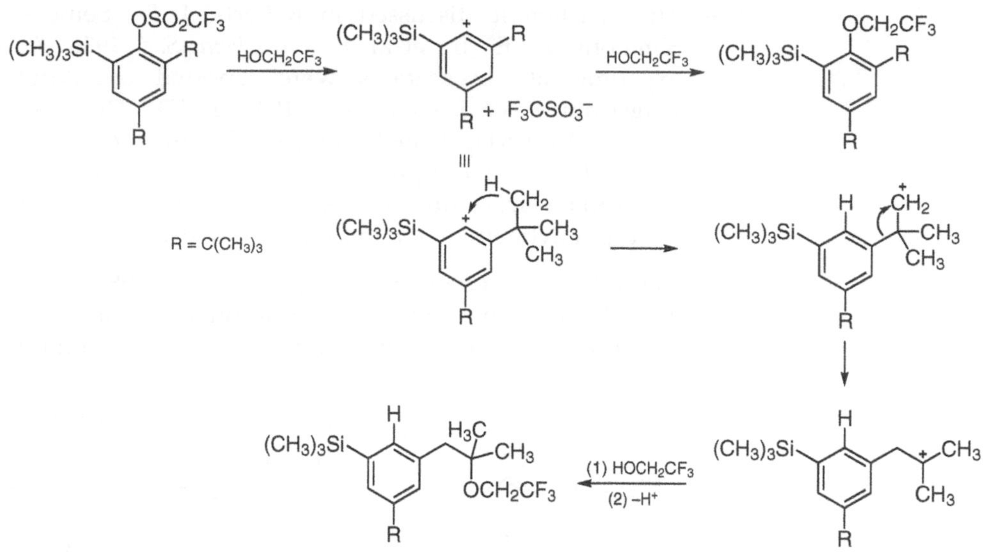

8.15. Grovenstein, E., Jr.; Kilby, D. C. *J. Am. Chem. Soc.* **1957**, *79*, 2972.

a. 3.97

b. The barrier for formation of the intermediate shown here is lower than that for conversion of the intermediate to product. Therefore, the C–H or C–D bond is broken in the rate-limiting step, and a 1° hydrogen isotope effect is observed. The intermediate is in equilibrium with reactants, so the rate constant for the reaction depends on the concentrations of both phenol and iodine.

8.16. The incorporation of protons from solvent was demonstrated by Bunnett, J. F.; Rauhut, M. M.; Knutson, D.; et al. *J. Am. Chem. Soc.* **1954**, *76*, **5755**, who found that the product contained deuterium when the reaction was run in D_2O/CH_3CH_2OD. Samuel, D. *J. Chem. Soc.* **1960**, 1318 studied the reaction in $H_2{}^{18}O/CH_3CH_2OH$ and found that the carboxylic acid group incorporated one ^{18}O atom. Rosenblum, M. *J. Am. Chem. Soc.* **1960**, *82*, **3796**, studied the reaction with reagents containing ^{15}N. For details of the information obtained from these mechanistic studies, see these and other references cited in the chapter.

8.17. If the mechanism shown operates in this reaction, it should be possible to detect *m*-chlorotrifluoromethylbenzene in the reaction mixture. If such a mechanism were to operate in general, then isotopic labeling should show formation of, for example, aniline-3-^{14}C from reaction of chlorobenzene-1-^{14}C.

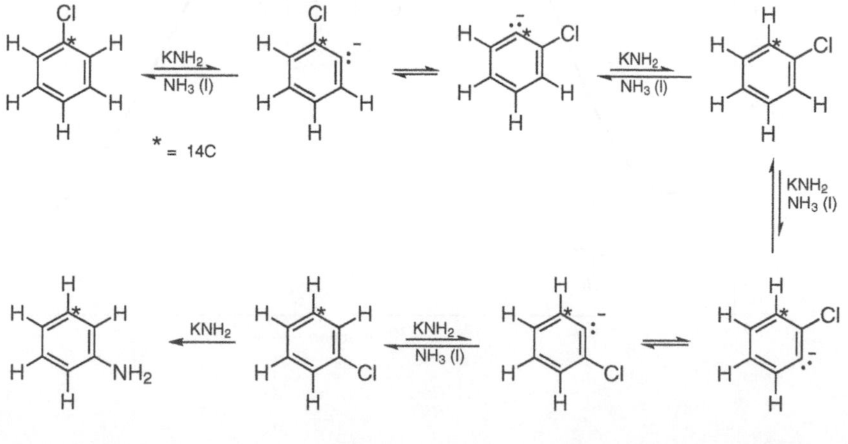

8.18. This reaction is discussed in Roberts, J. D.; Semenow, D. A.; Simmons, H. E., Jr.; et al. *J. Am. Chem. Soc.* **1956**, *78*, 601. The experimental observations were reported by Wright, R. E.; Bergstrom, F. W. *J. Org. Chem.* **1936**, *1*, 179. The observation of catalysis by KNH_2 strongly suggests that the mechanism is not an S_NAr substitution but proceeds, instead, by a benzyne mechanism. The KNH_2 reacts with chlorobenzene to form benzyne, which adds the triphenylmethide ion to form tetraphenylmethane.

8.19. Katritzky, A. R.; Laurenzo, K. S. *J. Org. Chem.* **1986**, *51*, 5039. This is an S_NAr reaction in which the Meisenheimer complex undergoes base-promoted ß-elimination to form a tautomer of the product.

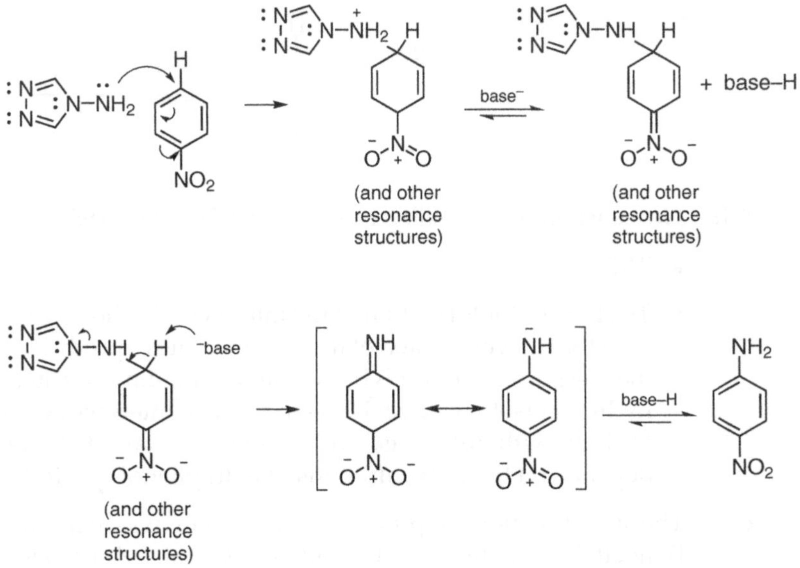
base⁻
+ base–H
(and other resonance structures)
(and other resonance structures)
base
(and other resonance structures)
base–H

8.20. Bunnett, J. F.; Garbisch, E. W., Jr.; Pruitt, K. M. *J. Am. Chem. Soc.* **1957**, *79*, 385. The lack of an element effect indicates that the reaction occurs in two steps: rate-limiting formation of a Meisenheimer complex, followed by elimination of the bromide ion.

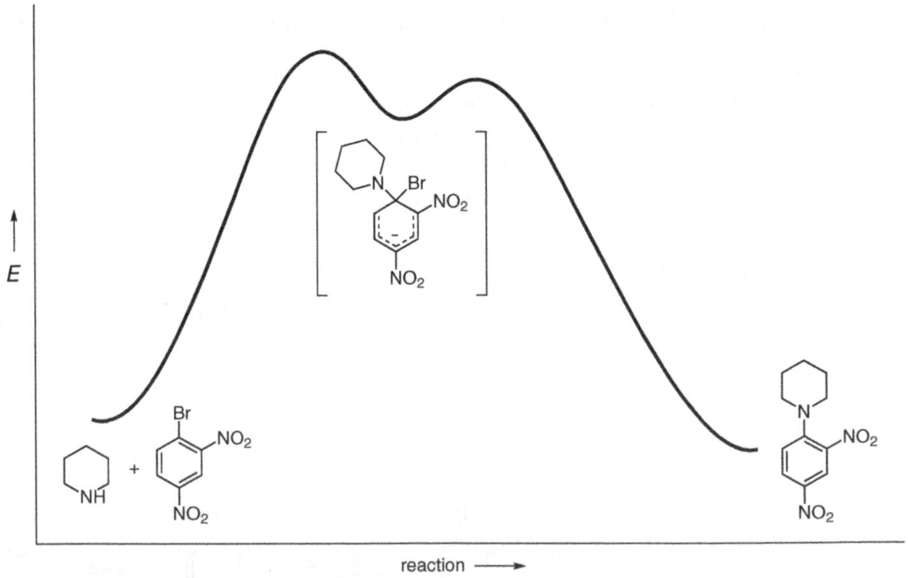
E
reaction

8.21. For a general discussion, see Roberts, J. D.; Vaughan, C. W.; Carlsmith, L. A.; et al. *J. Am. Chem. Soc.* **1956**, *78*, 611. Only one benzyne can be produced. Addition of amide ion to that benzyne gives two phenyl carbanions. The slight preference for meta product may result from a steric interaction between the amino group and an ortho methyl group.

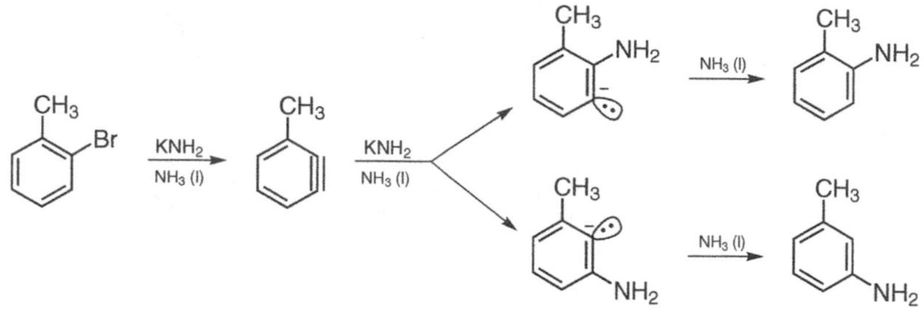

8.22. Roberts, J. D.; Semenow, D. A.; Simmons, H. E., Jr.; et al. *J. Am. Chem. Soc.* **1956**, *78*, 601.

a. These compounds are unreactive by the benzyne mechanism because in each case there is not a hydrogen atom ortho to the halogen atom.

b. The lack of reaction of these compounds alone does not establish the benzyne mechanism because in each case the halogen atom is sterically hindered by the ortho substituents. This steric hindrance could decrease the rate of any reaction step involving addition of a nucleophile to the carbon atom bearing the halogen atom or to an adjacent atom.

8.23. The figure here shows formation of the ortho product. The mechanism for formation of the para product is similar. See Makosza, M.; Winiarski, J. *J. Org. Chem.* **1984**, *49*, 1494; Makosza, M.; Ludwiczak, S. *J. Org. Chem.* **1984**, *49*, 4562; Makosza, M.; Winiarski, J. *Acc. Chem. Res.* **1987**, *20*, 282.

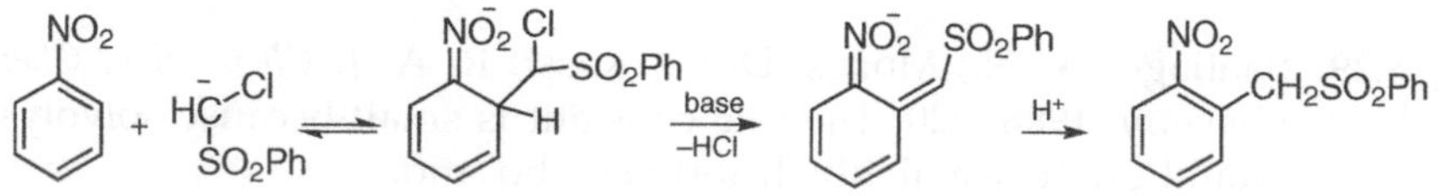

8.24. Malnar, I.; Jurić, S.; Vrček, V.; et al. *J. Org. Chem.* **2002**, *67*, 1490. $\Delta H^{\ddagger}$ is lower and $\Delta S^{\ddagger}$ is more negative for **110** because the double bond offers nucleophilic assistance to the departure of the chloride.

8.25. Zubkov, F. I.; Nikitina, E. V.; Kouznetsov, V. V.; et al. *Eur. J. Org. Chem.* **2004**, 5064.

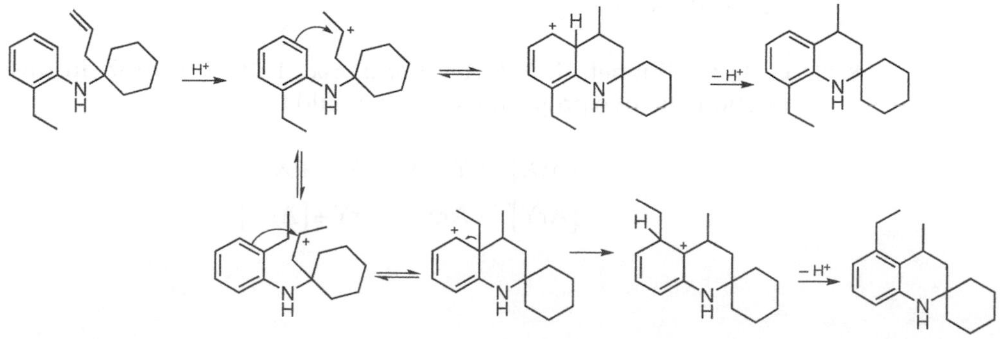

8.26. The first step is enolate formation, followed by product formation through S_Ni' reaction involving either the carbon or the oxygen of the enolate acting as nucleophile. Lovchik, M. A.; Goeke, A.; Fráter, G. *J. Org. Chem.* **2007**, *72*, 2427.

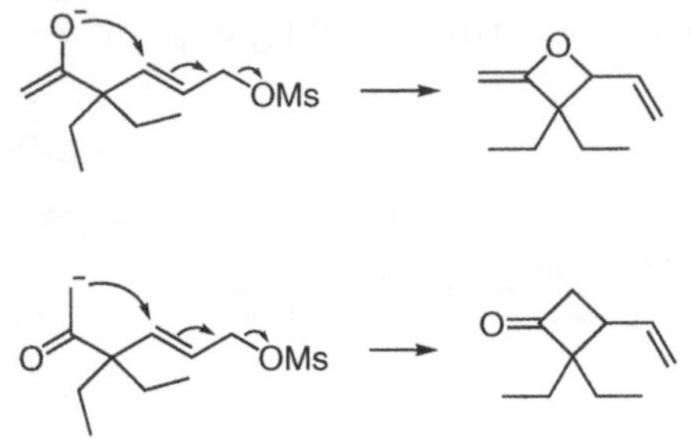

8.27. Pan, J.; Kampf, J. W.; Ashe, A. J., III. *Org. Lett.* **2007**, *9*, 679.

8.28. Sakamoto, Y.; Tamegai, K.; Nakata, T. *Org. Lett.* **2002**, *4*, 675. Pathways to the products are suggested by the arrows labeled a, b, and c in the scheme below.

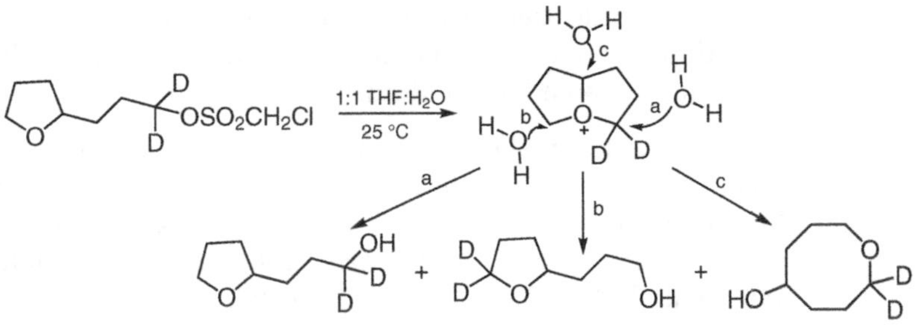

8.29. Santiago, A. N.; Morris, D. G.; Rossi, R. A. *J. Chem. Soc. Chem. Commun.* **1988**, 220. The rate constant is small because solvolysis would produce a highly distorted carbocation.

$\ln\frac{[R]}{[R_0]} = -kt$. At the half-life, $\ln\frac{[0.5]}{[1]} = -kt$, where $k = 2.7 \times 10^{-20}\,s^{-1}$.

Solving for t,

$$-\frac{\ln(0.5)}{2.7 \times 10^{-20}\,s^{-1}} = t = 2.57 \times 10^{19}\,s = 8.14 \times 10^{11}\text{ years.}$$

8.30. Galli, C.; Bunnett, J. F. *J. Am. Chem. Soc.* **1979**, *101*, 6137. In this reaction, the two propagation steps would be

$$[ArX]^{\cdot-} + Y^- \rightarrow [ArY]^{\cdot-} + X^-$$
$$[ArY]^{\cdot-} + ArX \rightarrow ArY + [ArX]^{\cdot-}$$

For discussions of this mechanism, see (a) Rossi, R. A.; Palacios, S. M. *Tetrahedron* **1993**, *49*, 4485; (b) Savéant, J.-M. *Tetrahedron* **1994**, *50*, 10117.

8.31. S_N2'. Marshall, J. A.; Bennett, C. E. *J. Org. Chem.* **1994**, *59*, 6110.

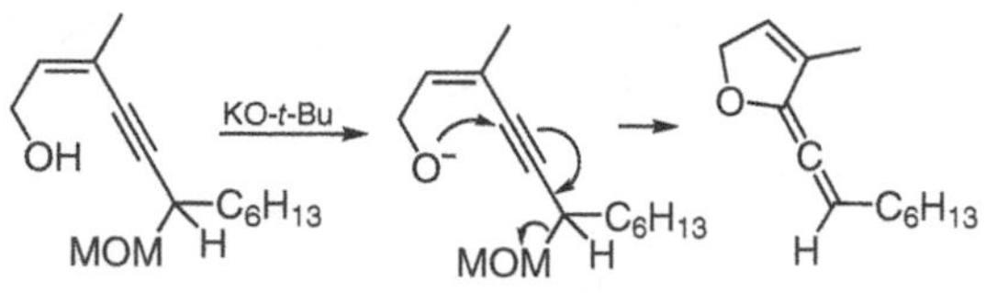

8.32. Mascal, M.; Hafezi, N.; Toney, M. D. *J. Am. Chem. Soc.* **2010**, *132*, 10662. There is no apparent S_N1 reaction, so the azide product must be formed by the S_N2 reaction of azide ion on a 3° carbon center.

8.33. Tsuji, Y.; Hara, D.; Hagimoto, R.; et al. *J. Org. Chem.* **2011**, *76*, 9568. Tsuji, Y.; Richard, J. P. *J. Phys. Org. Chem.* **2016**, *29*, 557. The first step is loss of tosylate involving anchimeric assistance of the phenyl ring to form a phenonium ion. Then attack of a nucleophile on either methylene of the three-membered ring leads to product with the isotopic label equally distributed on the product. The mechanism for just one of the products is shown here.

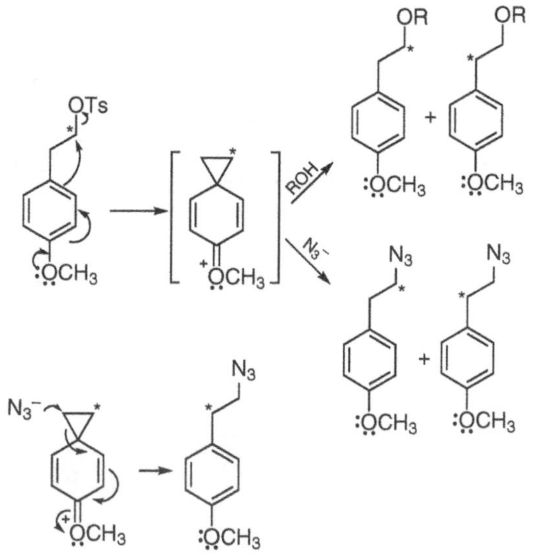

8.34. Dolbier, W. R., Jr.; Cornett, E.; Martinez, H.; et al. *J. Org. Chem.* **2011**, *76*, 3450. The proposed mechanism involves halide-assisted opening of an acylium ion to a ketene that is then protonated to form the acylium ion responsible for product formation.

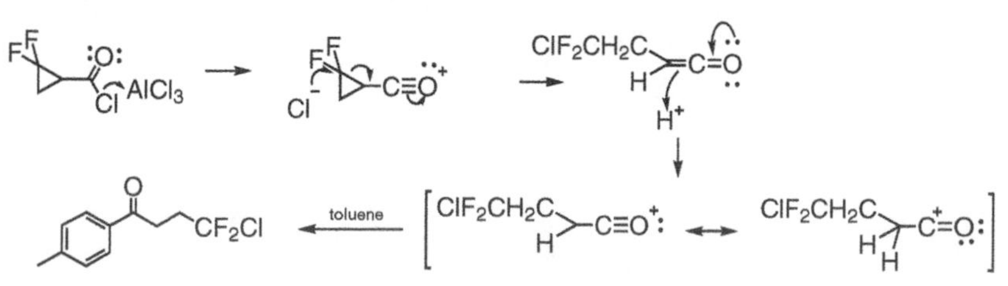

8.35. **a.** Intramolecular S_NAr reaction: Levy, A. A.; Rains, H. C.; Smiles, S. *J. Chem. Soc.* **1931**, 3264. For an extensive discussion, seeTruce, W. E.; Kreider, E. M.; Brand, W. W. *Org. React.* **1970**, *18*, 99.

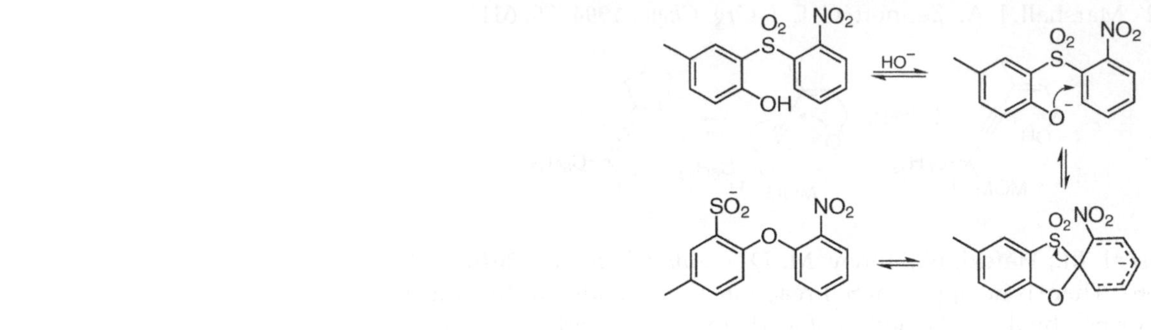

b. The proposed mechanism features an S_N2' reaction. Novi, M.; Dell'Erba, C.; Sancassan, F. *J. Chem. Soc. Perkin Trans. 1* **1983**, 1145.

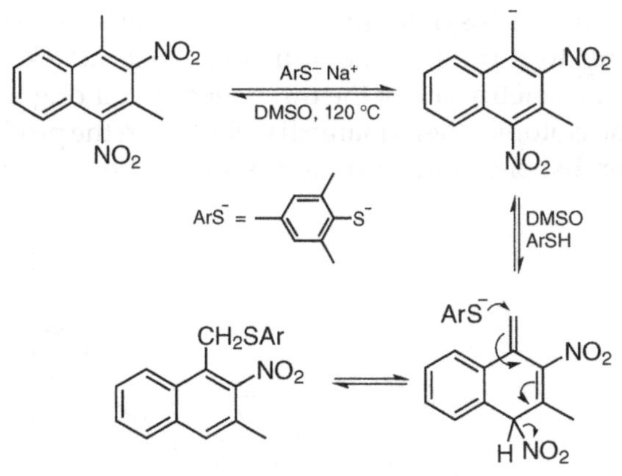

ArS⁻ Na⁺

DMSO, 120 °C

DMSO

ArSH

CHAPTER 9

Elimination Reactions

9.1. For discussions of these results, see pp. 2107 and 2101, respectively, of Dhar, M. L.; Hughes, E. D.; Ingold, C. K.; et al. *J. Chem. Soc.* **1948**, 2093. The first two reactions show Saytzeff orientation, while the second two show Hofmann orientation. In each case, however, the additional methyl gives a greater statistical basis for 1-alkene formation.

9.2. Cristol, S. J.; Hause, N. L. *J. Am. Chem. Soc.* **1952**, *74*, 2193. H and Cl can lie in the same plane in **9**, but in **10** they cannot.

H and Cl are syn-periplanar

H and Cl are neither syn-periplanar nor anti-periplanar

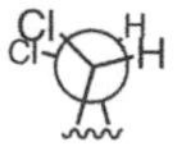

9 **10**

9.3. Kibby, C. L.; Lande, S. S.; Hall, W. K. *J. Am. Chem. Soc.* **1972**, *94*, 214. Syn elimination from *threo*-2-butanol-3-d_1 would give labeled *cis*-2-butene and unlabeled *trans*-2-butene. Anti elimination from *threo*-2-butanol-3-d_1 would give labeled *trans*-2-butene and unlabeled *cis*-2-butene. Syn elimination from *erythro*-2-butanol-3-d_1 would give labeled *trans*-2-butene and unlabeled *cis*-2-butene. Anti elimination from *erythro*-2-butanol-3-d_1 would give labeled *cis*-2-butene and unlabeled *trans*-2-butene. The reactions are illustrated (using Maehr notation) for the threo diastereomer in the figure shown here.

Solutions Manual for Perspectives on Structure and Mechanism in Organic Chemistry, Third Edition.
Felix A. Carroll.

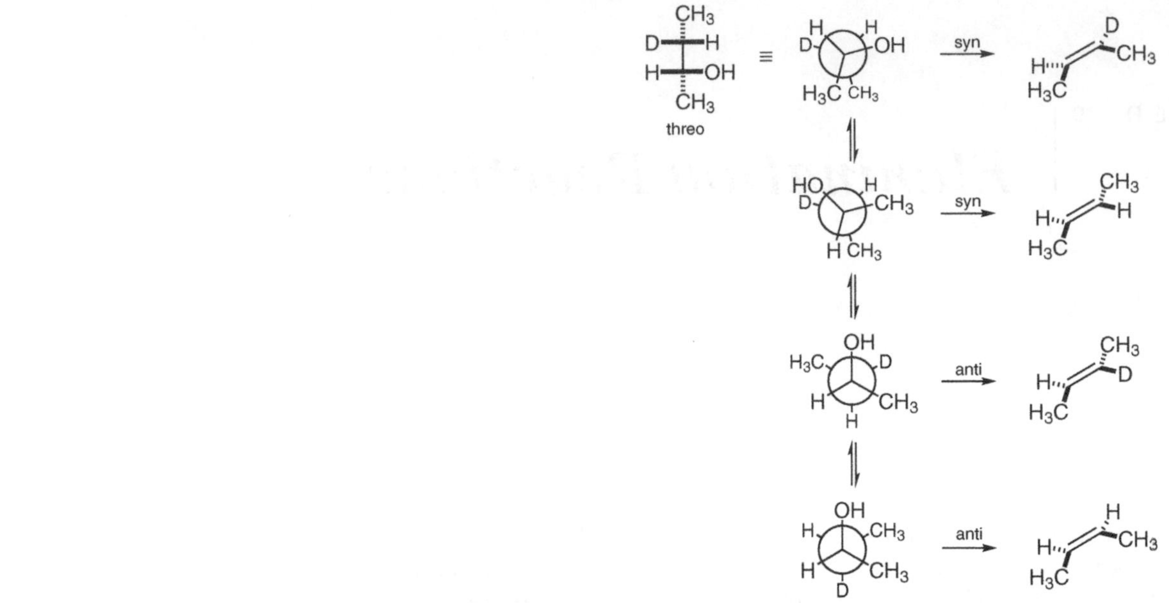

threo

syn

syn

anti

anti

9.4. Searles, S.; Gortatowski, M. J. *J. Am. Chem. Soc.* **1953**, *75*, 3030 proposed that the reaction occurs by a two-step mechanism in which the rate-limiting step is either an intramolecular S_N2 reaction or a fragmentation of an alkoxide ion.

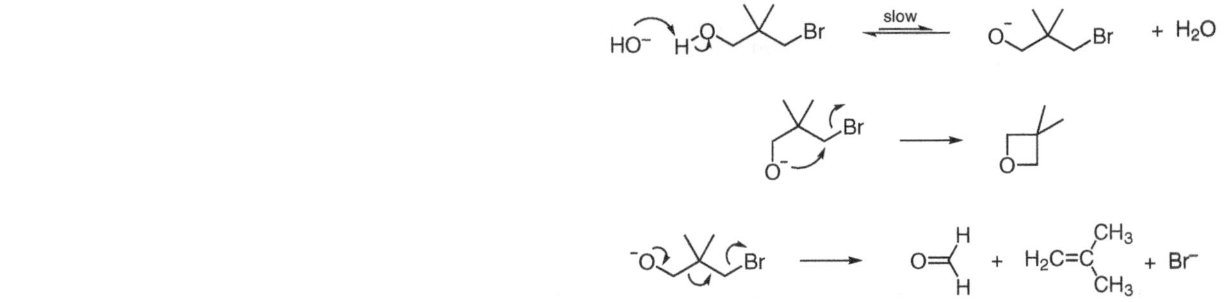

slow

$+ H_2O$

$+ H_2C{=}C(CH_3)_2 + Br^-$

9.5. Grovenstein, E., Jr.; Lee, D. E. *J. Am. Chem. Soc.* **1953**, *75*, 2639; Cristol, S. J.; Norris, W. P. *J. Am. Chem. Soc.* **1953**, *75*, 2645. The elimination in acetone is a concerted anti elimination from the carboxylate ion.

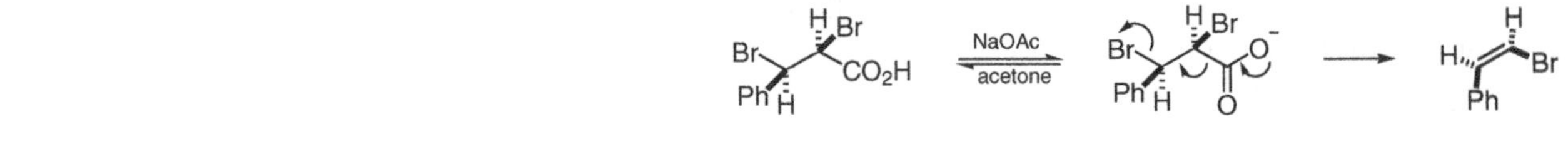

NaOAc

acetone

In a more polar solvent, ionization is competitive with elimination. The resulting carbocation can undergo conformational change before elimination of CO_2, so the more stable product is obtained.

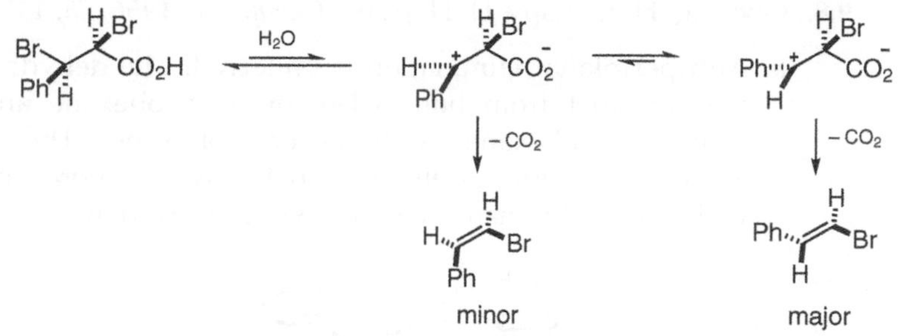

9.6. Ölwegård, M.; Ahlberg, P. *J. Chem. Soc. Chem. Commun.* **1989**, 1279; 1990, 788. Compounds **56** and **57** should give different distributions of deuterium-containing (*E*)- and (*Z*)-ethylideneindene, thus allowing determination of the relative rate constants for syn and anti elimination. The figure here uses the Maehr notation.

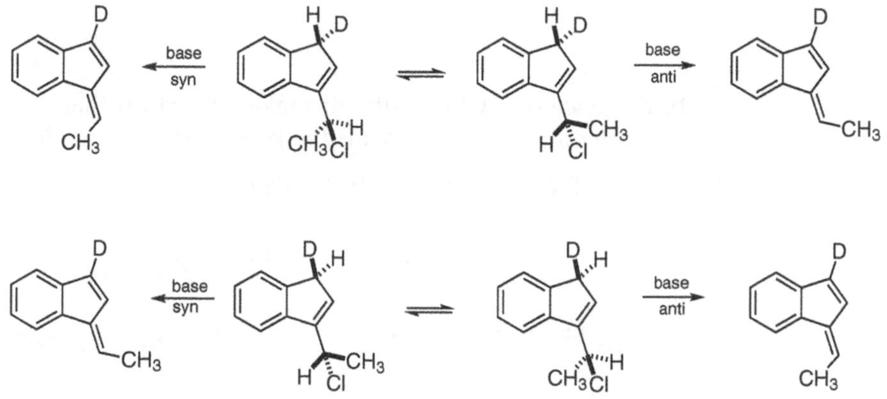

9.7. Kurtz, R. R.; Houser, D. J. *J. Org. Chem.* **1981**, *46*, 202.

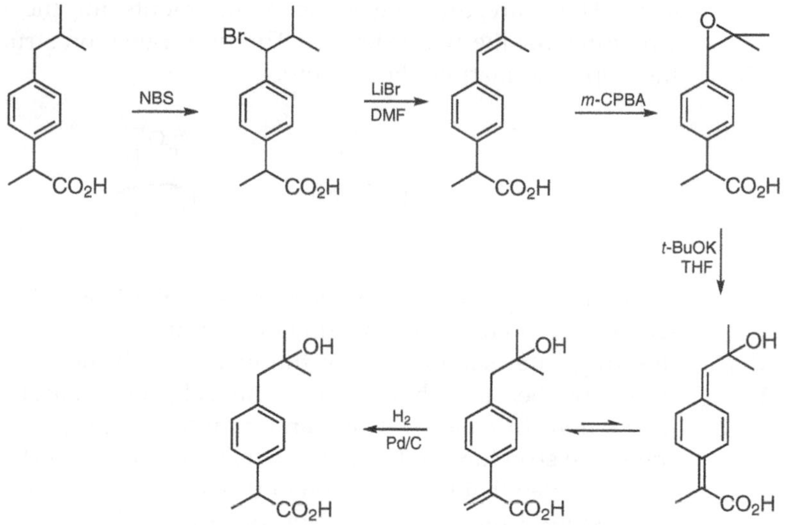

9.8. Goering, H. L.; Espy, H. H. *J. Am. Chem. Soc.* **1956**, *78*, 1454.

a. Anti periplanar elimination is preferred. The dehydrochlorination product from both 1,1-dichlorocyclohexane and *cis*-1,2-dichlorocyclohexane is 1-chlorocyclohexene. The trans 1,2 isomer can undergo elimination to give 3-chlorocyclohexene, which then eliminates to give 1,3-cyclohexadiene.

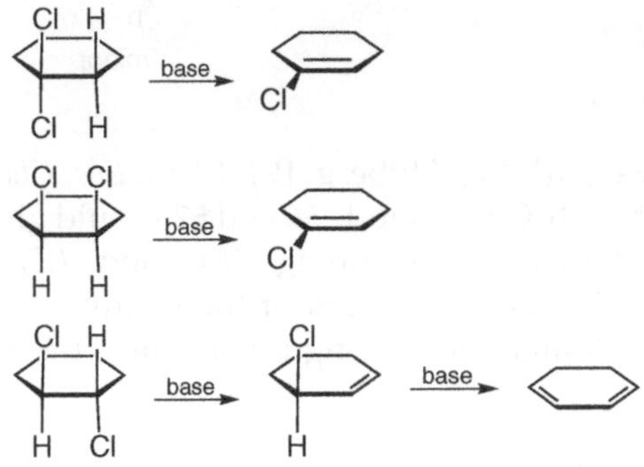

Cl H Cl H base Cl

Cl Cl H H base Cl

Cl H H Cl base Cl H base

b. Presence of a Cl substituent makes the H on that same carbon atom more acidic, so that is the preferred route of elimination.

9.9. Anh, N. T. *Chem. Commun.* **1968**, 1089.

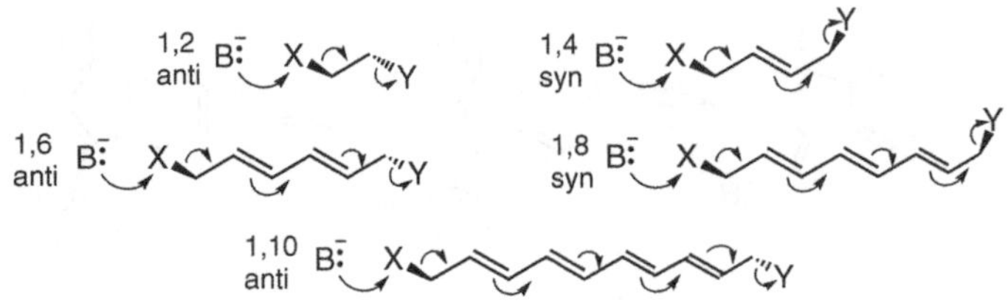

1,2 anti B: X Y

1,4 syn B: X Y

1,6 anti B: X Y

1,8 syn B: X Y

1,10 anti B: X Y

9.10. Cope, A. C.; LeBel, N. A.; Lee, H.-H.; et al. *J. Am. Chem. Soc.* **1957**, *79*, 4720. There are differing steric requirements for the transition structures for the two reactions. The two transition structures for the Cope elimination, shown here,

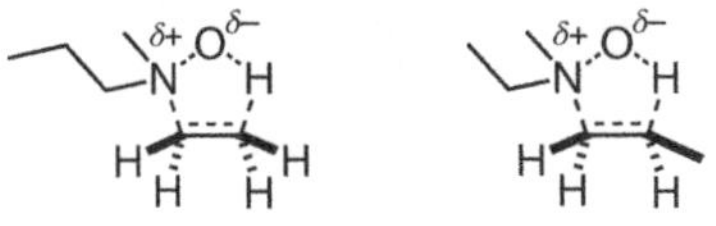

δ+ N O δ− H H H H H

δ+ N O δ− H H H H

represent eclipsed conformations of the alkyl branch on which elimination occurs. Any lowering of the transition structure energy due to greater stabilization of the incipient alkene appears to be offset by the steric hindrance of an eclipsed methyl–hydrogen interaction. In contrast, Cope and co-workers proposed that the transition structure for formation of ethene in equation 9.84 should be lower than that for formation of propene because the conformation required for anti elimination of propene has an unfavorable methyl–trialkylammonium gauche interaction.

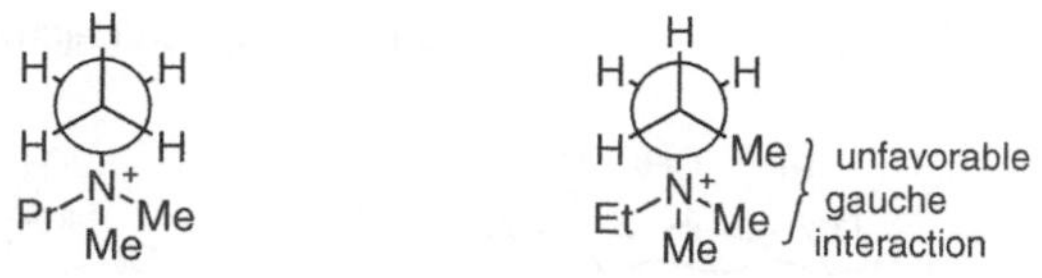

Conformation for elimination of ethene

unfavorable gauche interaction

Conformation for elimination of propene

9.11. Gandini, A.; Plesch, P. H. *J. Chem. Soc.* **1965**, 6019. The reaction was proposed to be a concerted elimination that should exhibit syn stereochemistry.

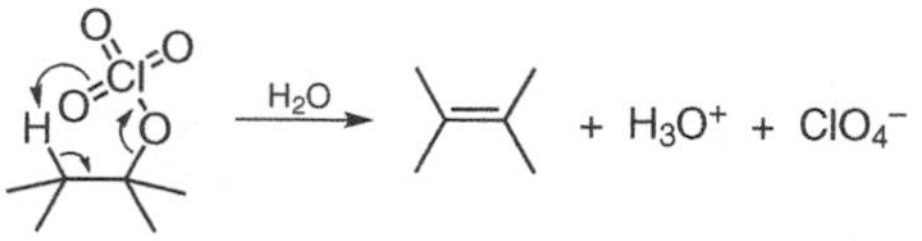

$\xrightarrow{H_2O}$ + H_3O^+ + ClO_4^-

9.12. O'Connor, G. L.; Nace, H. R. *J. Am. Chem. Soc.* **1953**, *75*, 2118.

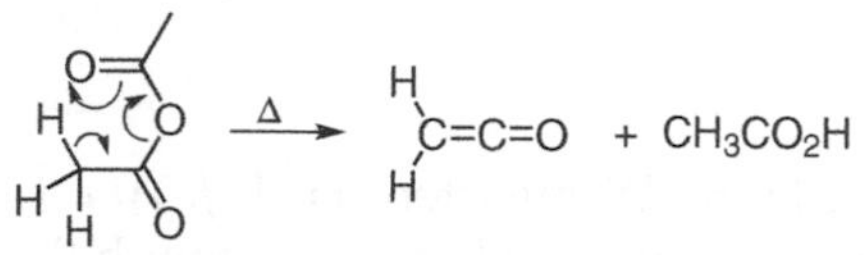

$\xrightarrow{\Delta}$ H₂C=C=O + CH_3CO_2H

9.13. Knözinger, H. in Patai, S., Ed. *The Chemistry of the Hydroxyl Group*, Part 2; Wiley-Interscience: London, 1971; pp. 660–661 and references therein. The alkoxide undergoes a fragmentation reaction.

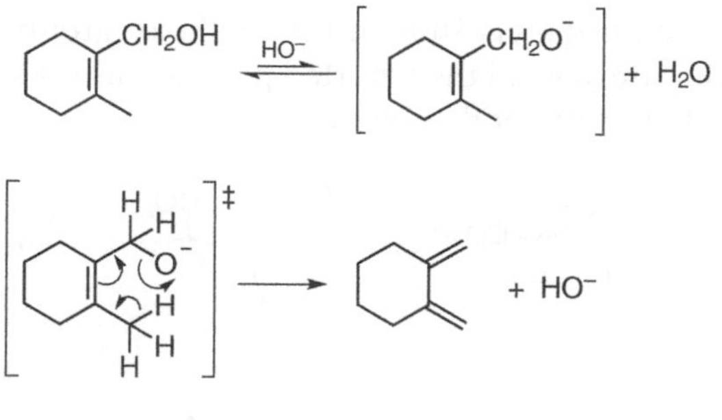

CH_2OH $\xrightarrow{HO^-}$ [CH_2O^-] + H_2O

[ ]‡ → + HO^-

A possible nonconcerted pathway could involve prior hydrogen abstraction by the alkoxide ion, leaving a CH_2^- group that can then undergo electron reorganization with expulsion of hydroxide ion. However, a very rough estimate of the difference in pK_a values of an alkane and an alcohol is about 25, so the equilibrium necessary for the nonconcerted mechanism would be very unfavorable.

9.14. Acharya, S. P.; Brown, H. C. *Chem. Commun.* **1968**, 305. The base should be as bulky as possible. The investigators used $(CH_3CH_2)_3CO^-K^+$ in $(CH_3CH_2)_3COH$ solution.

9.15. Barton, D. H. R.; Rosenfelder, W. J. *J. Chem. Soc.* **1951**, 1048. Both bromine atoms must be axial in 2ß,3α-2,3-dibromocholestane, so the anti-periplanar orientation for E2 elimination is feasible. In 3ß,4α-3,4-dibromocholestane, however, the two bromine atoms must be equatorial, so an anti-periplanar arrangement is not feasible.

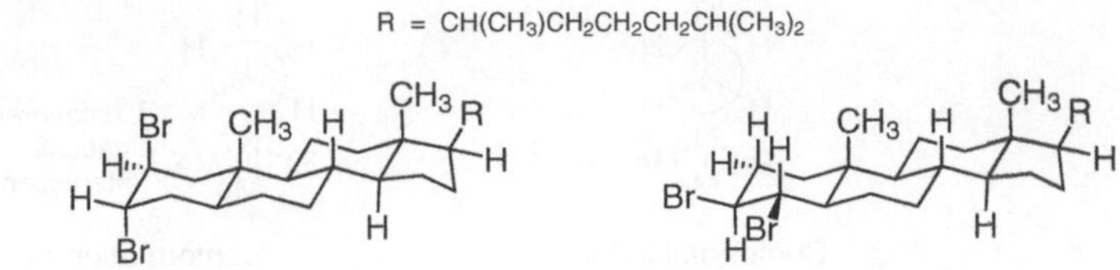

R = $CH(CH_3)CH_2CH_2CH_2CH(CH_3)_2$

2β,3α-2,3-dibromocholestane

3β,4α-3,4-dibromocholestane

9.16. Kashelikar, D. V.; Fanta, P. E. *J. Am. Chem. Soc.* **1960**, *82*, 4930. The proposed mechanism is a concerted reaction like that of the Chugaev or Cope reactions (Chapter 11).

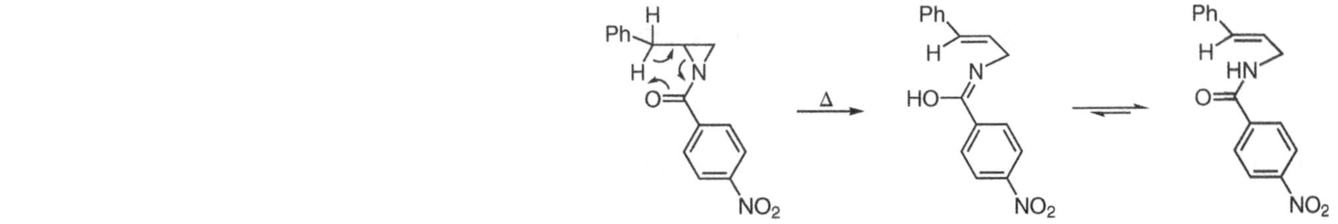

9.17. Noyce, D. S.; Weingarten, H. I. *J. Am. Chem. Soc.* **1957**, *79*, 3093. Interaction of the methoxy group with the acid chloride function is not feasible in the trans isomer, so the acid chloride is isolated.

With the cis isomer, the methoxy group reacts with the acid chloride to yield the reported products as a result of S_N2 or E2 attack by halide or another anion.

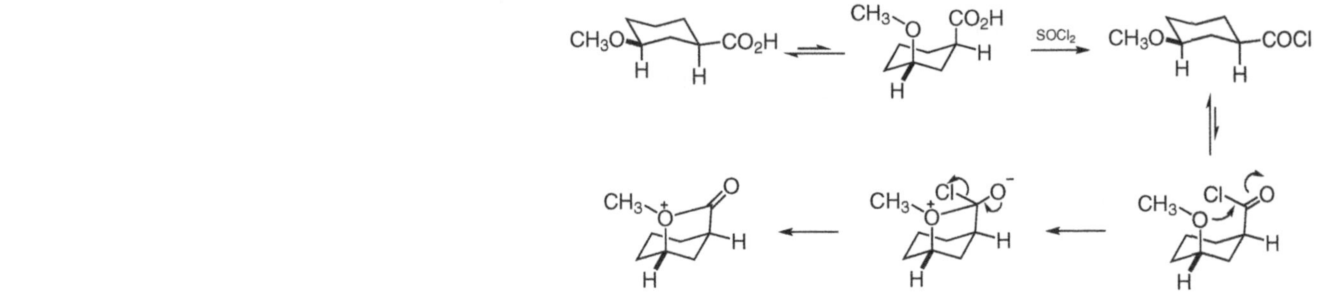

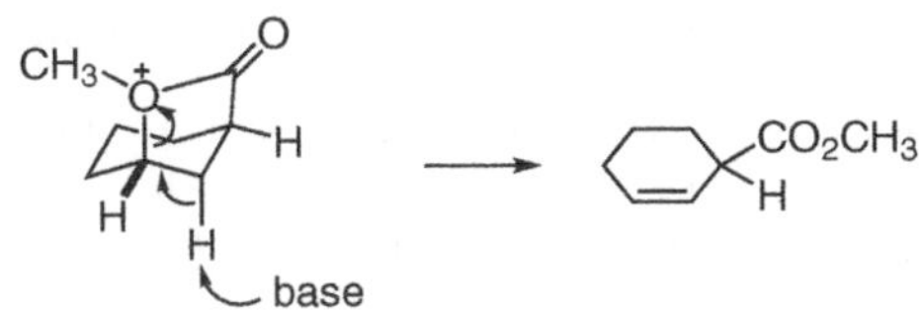

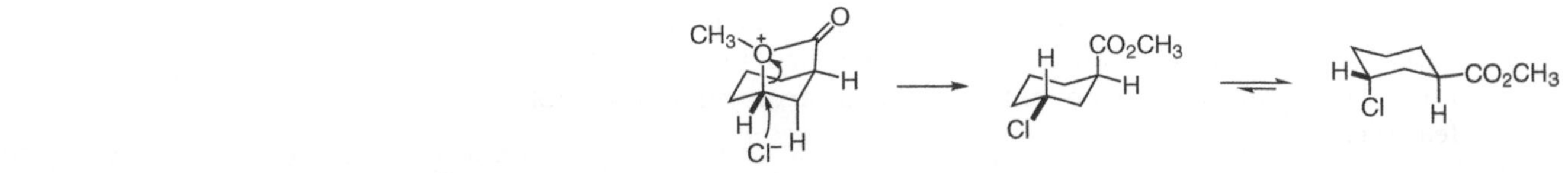

9.18. a. Alexander, E. R.; Mudrak, A. *J. Am. Chem. Soc.* **1950**, *72*, 1810. In the cis case, the hydrogen on the carbon atom with the phenyl group is not accessible in a six-membered transition structure. With the trans isomer, both hydrogens are accessible, although formation of the conjugated product is favored.

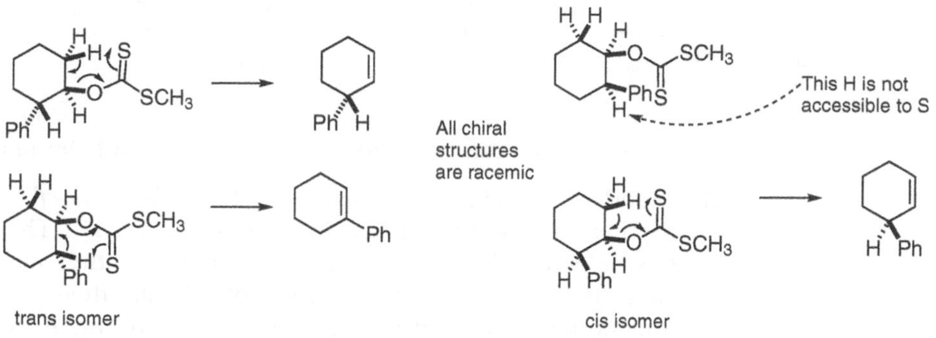

All chiral structures are racemic

This H is not accessible to S

trans isomer

cis isomer

b. Alexander, E. R.; Mudrak, A. *J. Am. Chem. Soc.* **1950**, *72*, 3194. The cis isomer has no sterically accessible hydrogen atom for a concerted reaction involving a six-membered transition structure.

c. Cope, A. C.; LeBel, N. A. *J. Am. Chem. Soc.* **1960**, *82*, 4656. The five-membered (H–C–C–N$^+$–O$^-$) ring transition structure shown in brackets is not sterically feasible in the smallest ring.

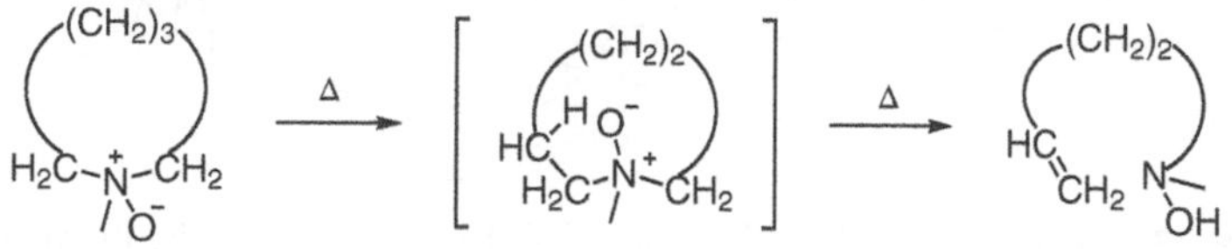

9.19. Kende, A. S. *Org. React.* **1960**, *11*, 261. The first step is a 1,3-elimination reaction.

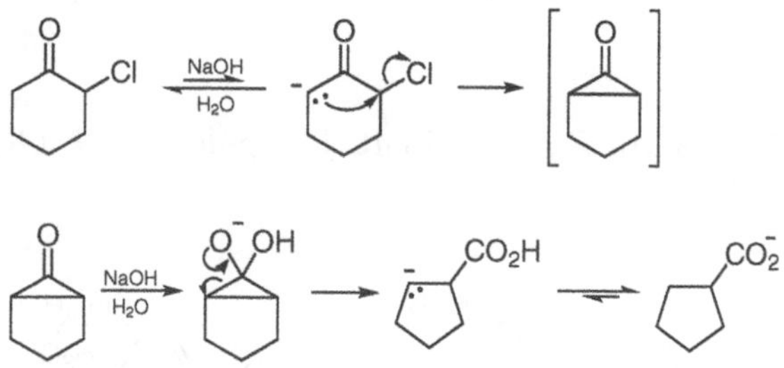

9.20. Curtin, D. Y.; Stolow, R. D.; Maya, W. *J. Am. Chem. Soc.* **1959**, *81*, 3330. The trans isomer cannot have an axial trimethylammonio group (because it would also have to have an axial *t*-butyl group), so it cannot undergo an E2 reaction. However, reaction of the substrate (at a methyl group) with *t*-butoxide can produce an S_N2 reaction.

$(CH_3)_3\overset{+}{N}$ H H $C(CH_3)_3$ ≡ CH_3 CH_3 N^+ CH_3 CH_3 CH_3 CH_3 *t*-BuO⁻ → CH_3 :N CH_3 CH_3 CH_3 CH_3 + *t*-BuOH

$(CH_3)_2N$ H H $C(CH_3)_3$ ≡

The cis isomer can undergo both substitution and elimination.

9.21. Saunders, W. H., Jr.; Cockerill, A. F. *Mechanisms of Elimination Reactions*; Wiley-Interscience: New York, 1973; p. 173. The former is less substituted, but the latter has *t*-butyl–methyl repulsion. If Saytzeff means giving the more stable alkene, then it is Saytzeff orientation. If Saytzeff means giving the more highly substituted alkene, then it is Hofmann orientation.

9.22. Lenoir, D.; Chiappe, C. *Chem. Eur. J.* **2003**, *9*, 1036 and references therein. The initial bromonium ion is too sterically hindered for the bromide ion to add, so bromide acts as a base instead:

Br_2, $ClCH_2CH_2Cl$, 25 °C → H H H Br^+ Br^- → H H H Br Br^- →

9.23. de Groot, F. M. H.; Loos, W. J.; Koekkoek, R.; et al. *J. Org. Chem.* **2001**, *66*, 8815.

H N PG O O Drug

9.24. Rappoport, Z.; Greenblatt, J.; Apeloig, Y. *J. Org. Chem.* **1979**, *44*, 3687.

OCH_3 HO H Br CH_3CO_2H, $-H_2O$ → $:\ddot{O}CH_3$ + Br → $\overset{+}{O}$–CH_3 H_2O Br → O Br

9.25. The Cope elimination is stereospecific, so the erythro reactant gives (*Z*)-2-phenyl-2-butene, while the threo reactant gives the *E* diastereomer. See Acevedo, O.; Jorgensen, W. L. *J. Am. Chem. Soc.* **2006**, *128*, 6141 and reference therein to DePuy, C. H.; King. R. W. *Chem. Rev.* **1960**, *60*, 431.

9.26. The 1 : 1 : 1 triplet on hydrogen bonded to C′ indicates that the deuterium atom at that position in the reactant remains on that carbon in the product. Therefore, the microorganism removed the hydrogen atom on C′ and the hydrogen atom on the adjacent carbon by a syn pathway. See Amir-Heidari, B.; Mickelfield, J. *J. Org. Chem.* **2007**, *72*, 8950.

9.27. de Groot, F. M. H.; Albrecht, C.; Koekkoek, R.; et al. *Angew. Chem. Int. Ed.* **2003**, *42*, 4490.

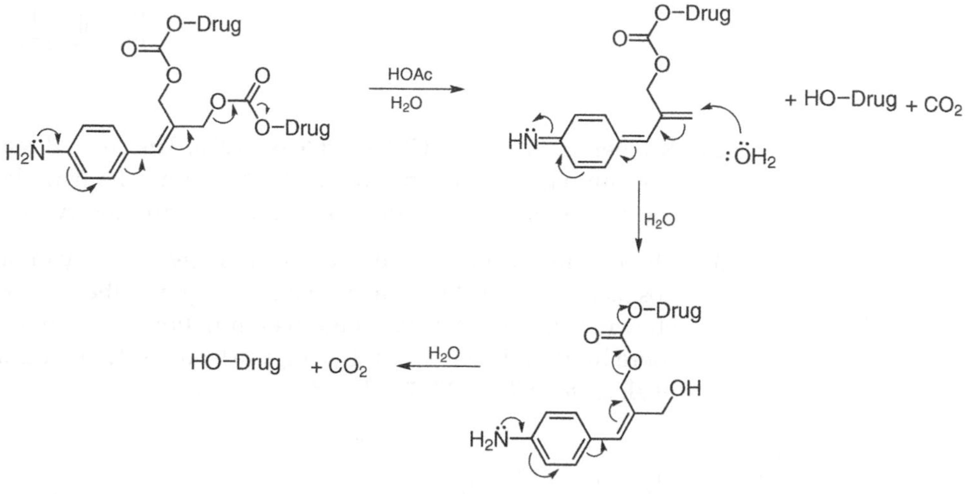

HOAc, H_2O → + HO–Drug + CO_2

H_2O

HO–Drug + CO_2 ← H_2O

9.28. **a.** This product is formed by a Cope elimination. Grainger, R. S.; Patel, A. *Chem. Commun.* **2003**, 1072.

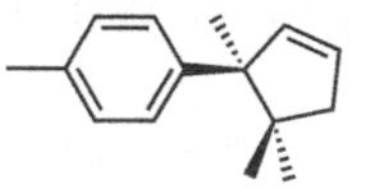

b. Honda, K.; Tabuchi, M.; Kurokawa, H.; et al. *J. Chem. Soc. Perkin Trans. 1* **2002**, 1387.

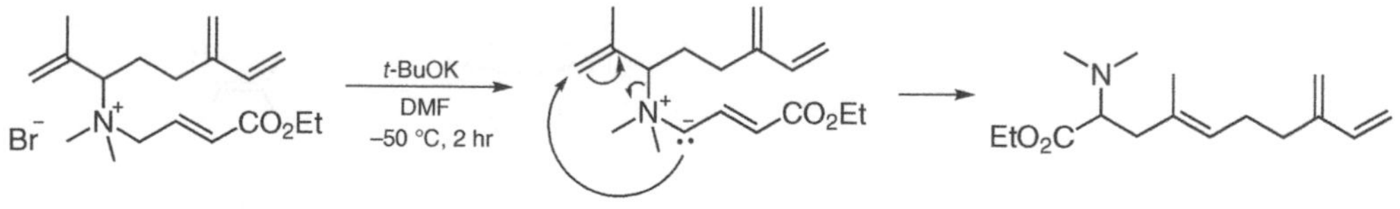

t-BuOK, DMF, –50 °C, 2 hr

c. Riether, D.; Mulzer, J. *Eur. J. Org. Chem.* **2003**, 30. Reactions are reductive animation (at the aldehyde function) to give the dimethylamine, oxidation to the *N*-oxide, and Cope elimination:

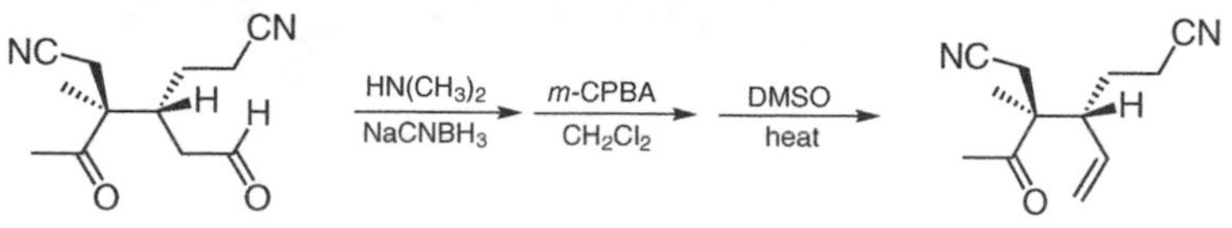

$HN(CH_3)_2$, $NaCNBH_3$ → *m*-CPBA, CH_2Cl_2 → DMSO, heat

d. Pye, P. J.; Rossen, K.; Weissman, S. A.; et al. *Chem. Eur. J.* **2002**, *8*, 1372. The product is formed by an S_N2 reaction followed by Hofmann elimination:

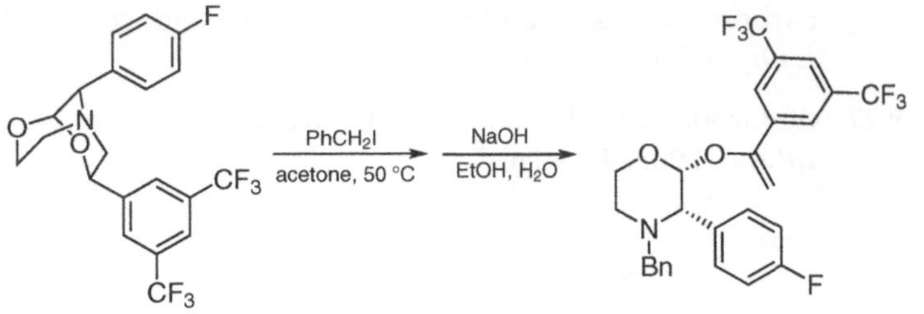

F
O N O
CF3
CF3
PhCH2I
acetone, 50 °C
NaOH
EtOH, H2O
F3C
CF3
O O
N
Bn
F

9.29. Kluiber, R. W. *J. Org. Chem.* **1966**, *61*, 1298. The second reaction is a 1,6-elimination, probably by an E1cb-like mechanism, because of the greater acidity of proton on carbon bearing the cyano group.

9.30. Hydroxide addition to the ester function gives an oxyanion that can abstract the H (or D) by a syn pathway, giving the stereochemistry shown. When R = *t*-butyl, the syn elimination pathway is sterically more hindered. See Mohrig, J. R.; Carlson, H. K.; Coughlin, J. M.; et al. *J. Org. Chem.* **2007**, *72*, 793.

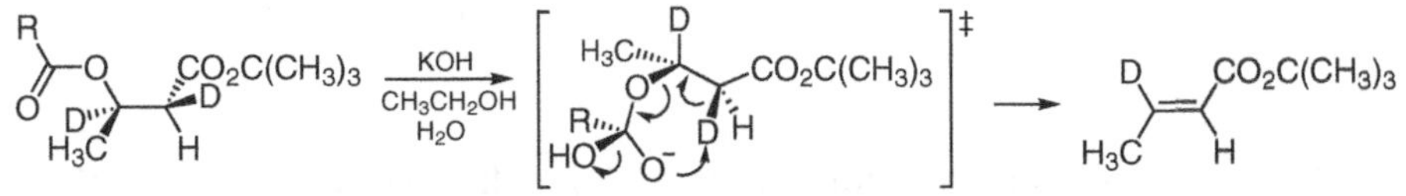

R
O O
D
H3C H
CO2C(CH3)3
D
KOH
CH3CH2OH
H2O
[D
H3C
CO2C(CH3)3
O
R
HO O⁻
D H]‡
→
D CO2C(CH3)3
H3C H

9.31. Acevedo, O.; Jorgensen, W. L. *J. Am. Chem. Soc.* **2006**, *128*, 6141.

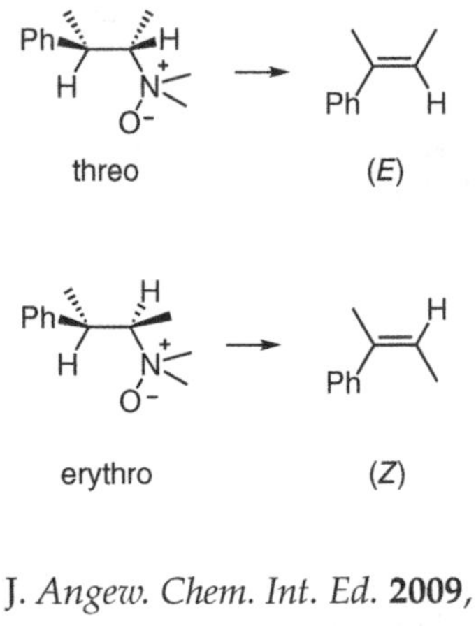

Ph H
H N+
O⁻
→
Ph H
threo (*E*)

H
Ph
H N+
O⁻
→
H
Ph
erythro (*Z*)

9.32. Prantz, K.; Mulzer, J. *Angew. Chem. Int. Ed.* **2009**, *48*, 5030.

R1
OMs
O R2
O
HO⁻
R1 OMs
O R2
OH O⁻

R1
OH
H
R2
− CO2
R1
HO2C O R2
H

9.33. Esselman, B. J.; Hofstetter, H.; Ellison, A. J.; et al. *J. Chem. Educ.* **2020**, *97*, 2280. Reaction in aqueous H_2SO_4 is an E1 reaction, while reaction in hot, basic 1-propanol is an E2 reaction. A reaction coordinate diagram for the E2 reaction would be similar to Figure 9.28. A reaction coordinate diagram for the E1 reaction is shown here. The $\delta\Delta G^{\ddagger}$ values for formation of the two products are different in E1 and E2 reactions. In addition, the products can be equilibrated by a protonation–deprotonation mechanism under acidic conditions but not under basic conditions.

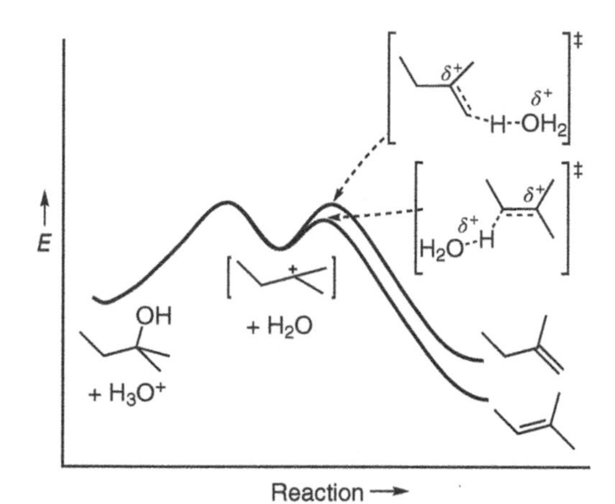

CHAPTER 10

Addition Reactions

10.1. Buckles, R. E.; Bader, J. M.; Thurmaier, R. J. *J. Org. Chem.* **1962**, *27*, 4523. The meso product is more stable, so it would be expected to be the major product if a reaction produces the more stable product regardless of reactant stereochemistry. Thus, investigating the reaction of both isomers is necessary to determine whether the reaction is stereospecific or merely stereoselective.

10.2. Buckles, R. E.; Forrester, J. L.; Burham, R. L.; et al. *J. Org. Chem.* **1960**, *25*, 24. The product is 2-bromo-3-chloro-3-phenylpropionic acid; the stereochemistry is presumed to be that from anti-addition. The regiochemistry is consistent with a mechanistic model in which BrCl adds as Br^+ and Cl^-, with Br^+ adding first to produce the more stable carbocation. (The figure below and several subsequent figures use the Maehr notation to indicate racemic compounds.)

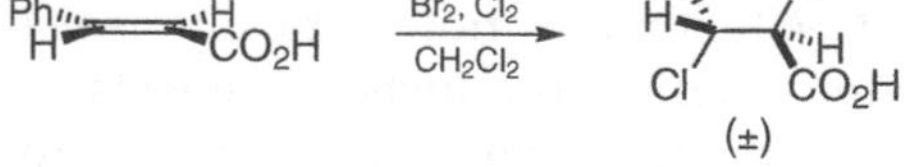

Solutions Manual for Perspectives on Structure and Mechanism in Organic Chemistry, Third Edition.
Felix A. Carroll.

10.3. Bellucci, G.; Bianchini, R.; Chiappe, C. *J. Org. Chem.* **1991**, *56*, 3067. CH_3CN adds as a nucleophile to the bromonium ion; subsequent addition of water, deprotonation, and tautomerization produces the product.

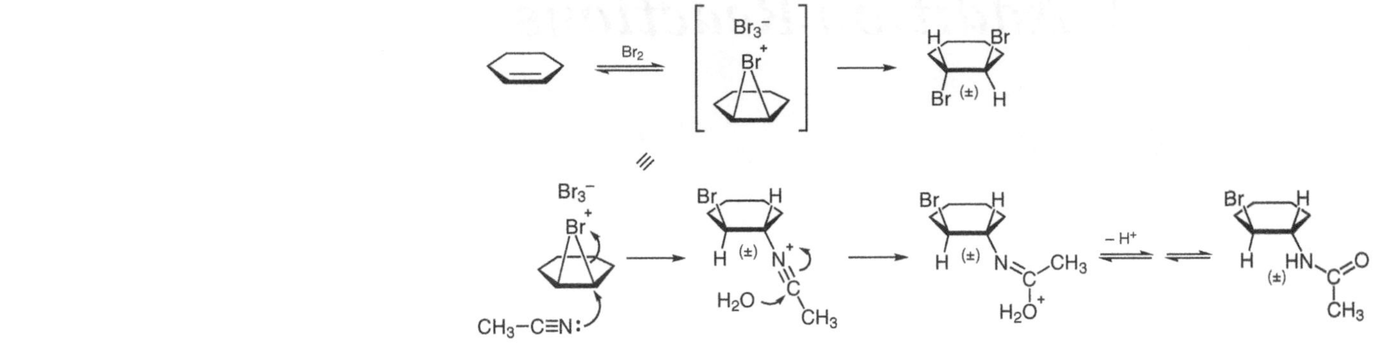

10.4. Bedoukian, P. Z. *J. Am. Chem. Soc.* **1944**, *66*, 1325. Acetylation of the enol of the reactant leads to the enol acetate (**A**), which adds bromine to form (**B**). Methanolysis leads to the formation of dimethyl acetal of α-bromophenylacetaldehyde (**C**).

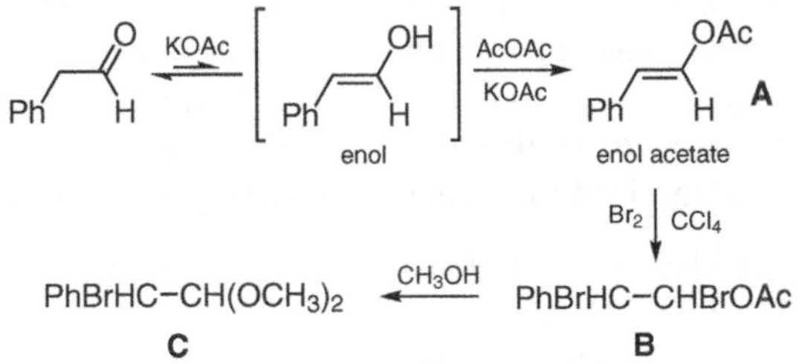

10.5. Tarbell, D. S.; Bartlett, P. D. *J. Am. Chem. Soc.* **1937**, *59*, 407. The lactone is formed from a chloronium ion or chlorocarbocation. It is unlikely that the chlorolactone is formed through intramolecular reaction of a chloronium ion intermediate because nucleophilic attack of the carboxylate group on the back side of a chloronium ion C–Cl bond is not sterically feasible. The observation of diastereomeric products suggests that closure of the chlorocarbocation must be faster than rotation about the carbon–carbon single bond. A mechanism for reaction of the (*E*)-diastereomer of the reactant is shown here, and the mechanism for the (*Z*)-diastereomer is analogous.

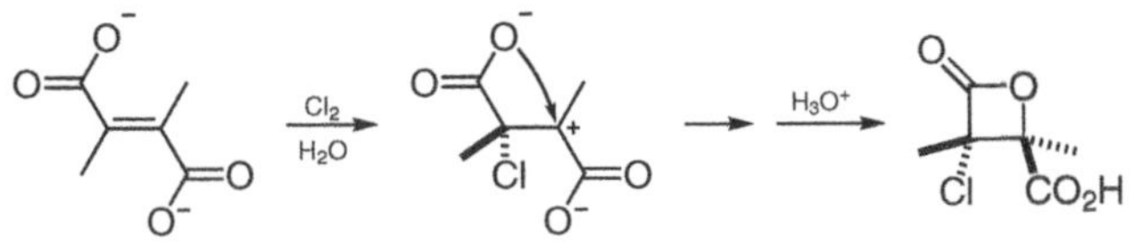

10.6. Rolston, J. H.; Yates, K. *J. Am. Chem. Soc.* **1969**, *91*, 1477. Dioxane acts as a nucleophile to produce an adduct (shown below), which can then undergo attack by Br^- after rotation about the carbon–carbon single bond.

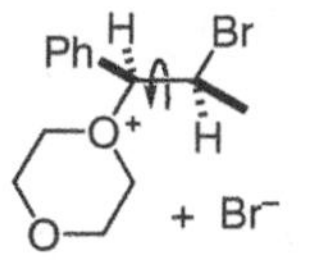

Ph H Br O+ H O + Br^-

10.7. van Tamelen, E. E.; Shamma, M. *J. Am. Chem. Soc.* **1954**, *76*, 2315 (parts a through d). For parts a through c, an intermediate iodinium ion reacts with a neighboring carboxylate group to form a 5- or 6-membered lactone ring. In part d, the iodinium ion does not react with the carboxylate group because formation of a 5- or 6-membered ring is not sterically feasible. Instead, a decarboxylation occurs to give a neutral compound.

a. CO_2H $\xrightarrow[\text{aq. NaHCO}_3]{I_2,\ KI}$ →

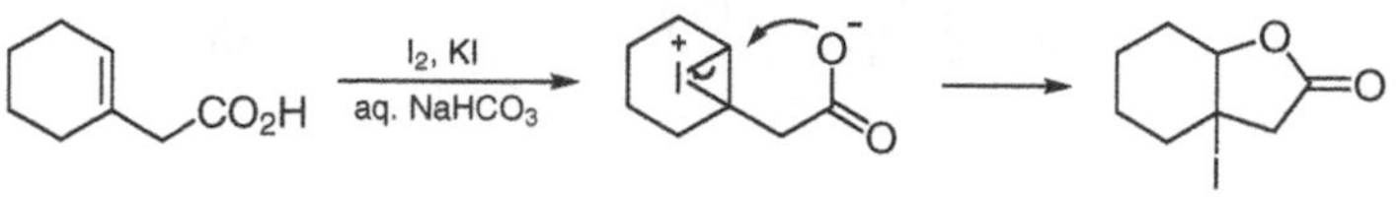

b. CO_2H $\xrightarrow[\text{aq. NaHCO}_3]{I_2,\ KI}$ →

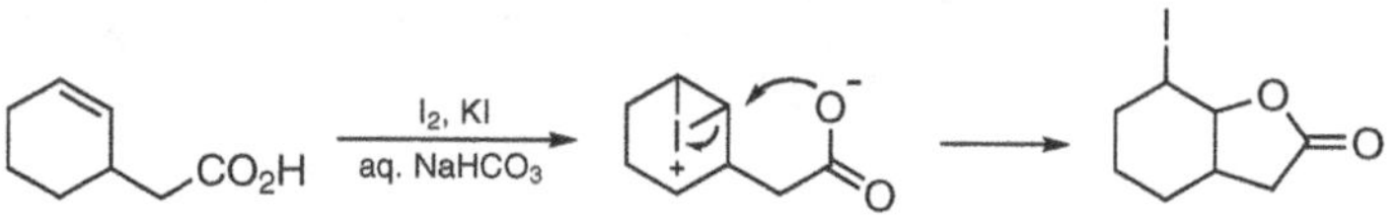

c. CO_2H $\xrightarrow[\text{aq. NaHCO}_3]{I_2,\ KI}$ → CH_2I

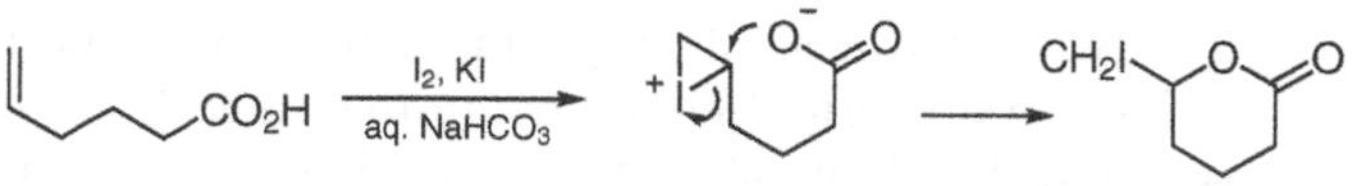

d. CO_2H $\xrightarrow[\text{aq. NaHCO}_3]{I_2,\ KI}$ → + CO_2 $\xrightarrow{H_2O}$ OH

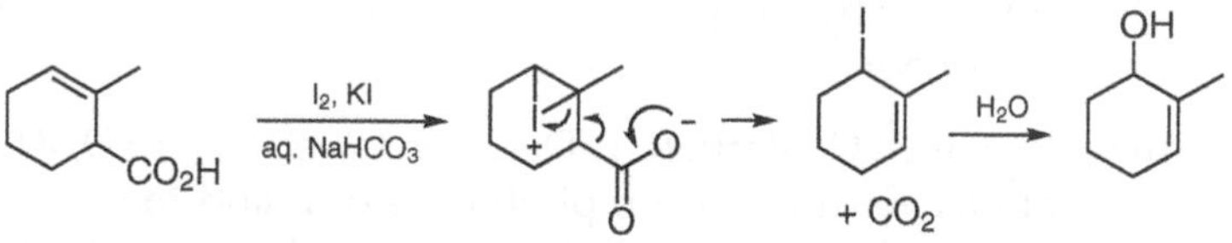

e. Butters, M.; Elliott, M. C.; Hill-Cousins, J.; et al. *Org. Lett.* **2007**, *9*, 3635. The authors discuss the regioselectivity of this and related reactions.

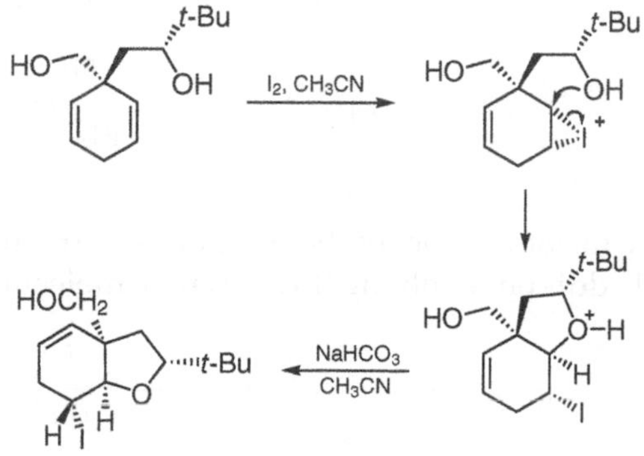

HO t-Bu OH $\xrightarrow{I_2,\ CH_3CN}$ HO t-Bu OH I+ ↓ HO t-Bu O+–H H I $\xrightarrow[CH_3CN]{NaHCO_3}$ $HOCH_2$ t-Bu O H I H

f. Collum, D. B.; McDonald, J. H., III.; Still, W. C. *J. Am. Chem. Soc.* **1980**, *102*, 2118. This is an iodolactonization reaction.

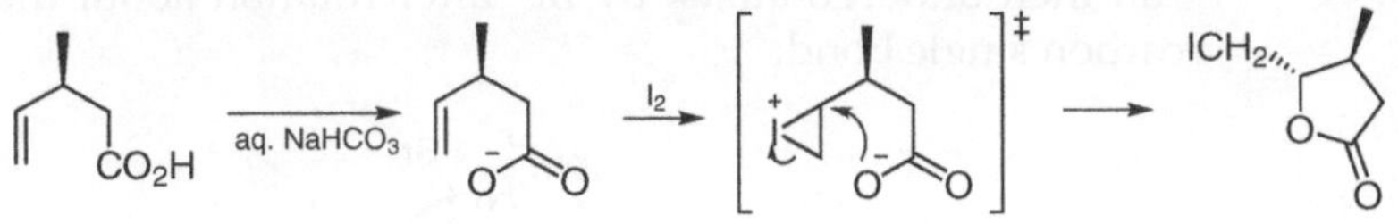

10.8. Peterson, P. E.; Tao, E. V. P. *J. Am. Chem. Soc.* **1964**, *86*, 4503. The observation of 5-chloro-2-deuterio-2-hexyl trifluoroacetate is consistent with the involvement of a 5-membered cyclic chloronium ion:

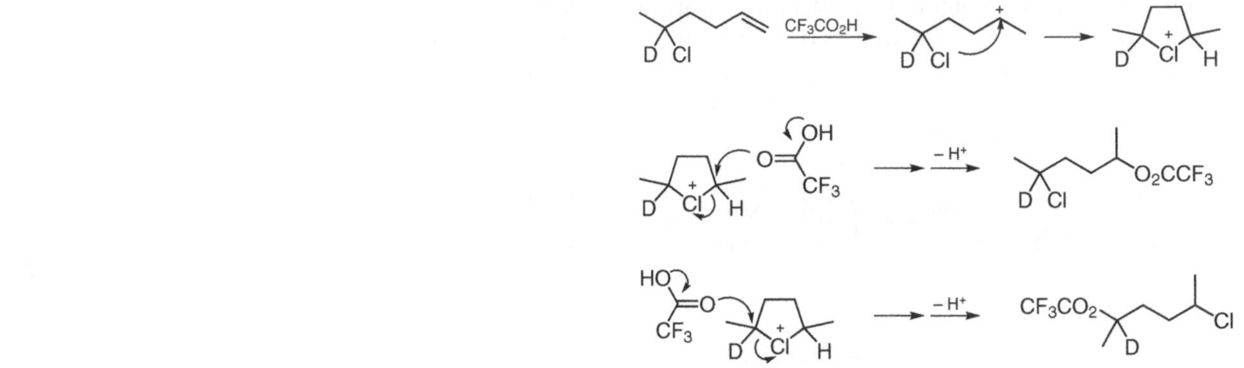

Since the chloronium ion should lead to essentially equal yields of the two products, it appears that about one-third of the 5-chloro-5-deuterio-2-hexyl trifluoroacetate is formed through addition of trifluoroacetic acid by ordinary Markovnikov addition (that is by a pathway not involving the chloronium ion intermediate).

10.9. Fraenkel, G.; Bartlett, P. D. *J. Am. Chem. Soc.* **1959**, *81*, 5582. Iodine atoms generated from photolysis of I_2 add to styrene to produce iodomethylbenzyl radicals that dimerize to produce the product.

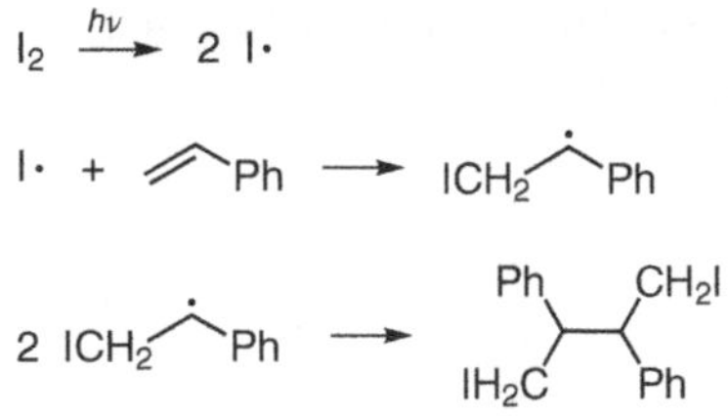

$$I_2 \xrightarrow{h\nu} 2\ I\cdot$$

The major product of the reaction is a structure suggesting a radical addition involving three styrene molecules.

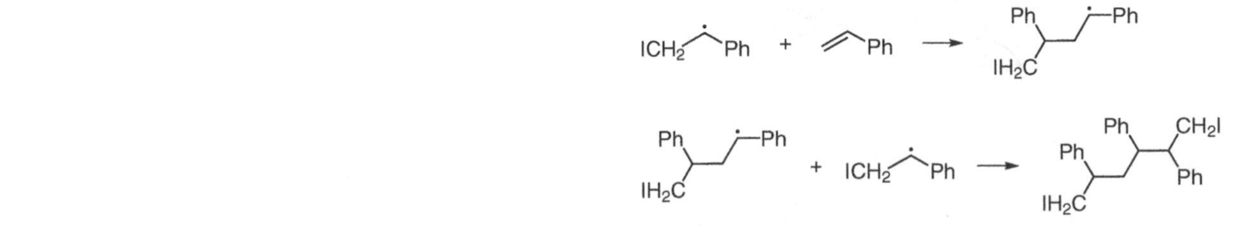

10.10. Okuyama, T.; Sakagami, T.; Fueno, T. *Tetrahedron* **1973**, *29*, 1503. See also the discussion in Chwang, W. K.; Knittel, P.; Koshy, K. M.; et al. *J. Am. Chem. Soc.* **1977**, *99*, 3395. Protonation of the terminal methylene group produces a resonance-stabilized carbocation that can undergo attack by water (by either or both of the routes shown) to produce the ketone.

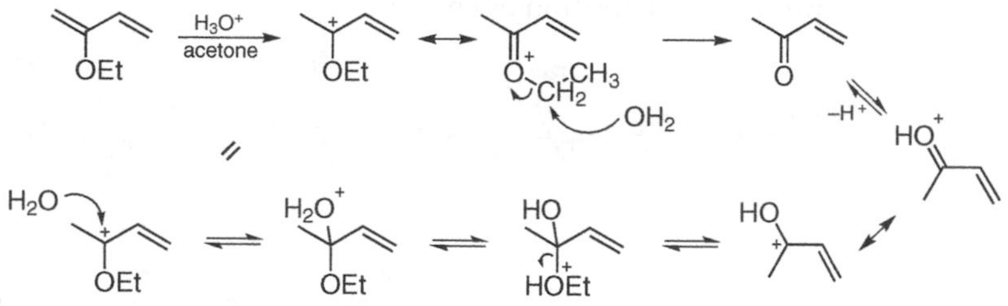

10.11. Bellucci, G.; Bianchini, R.; Vecchiani, S. *J. Org. Chem.* **1987**, *52*, 3355. The first two products are formed by attack of Br^- (or Br_3^-) on a bromonium ion. The last two products arise from attack of Br^- (or Br_3^-) on an intermediate resulting from neighboring group participation of the ester function.

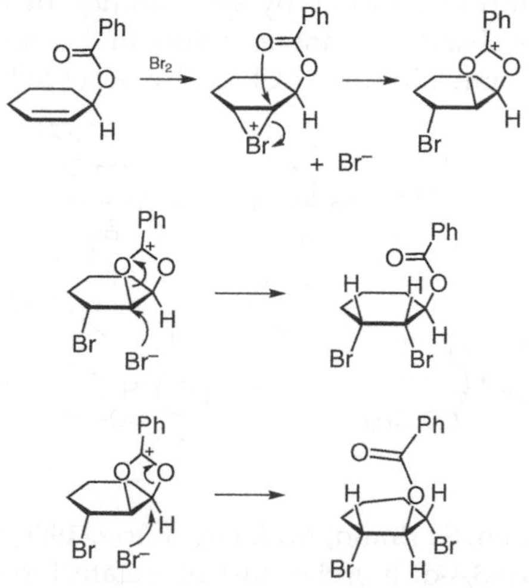

10.12. Fahey, R. C.; Schneider, H.-J. *J. Am. Chem. Soc.* **1968**, *90*, 4429. The product distributions correlate with the bridging ability of the adding electrophilic atom. Those atoms that bridge well (such as Br) lead to predominantly anti addition. Those that bridge poorly or not all (such as H) lead to open cations that can rotate before addition of the nucleophile, so a mixture of syn and anti addition is observed.

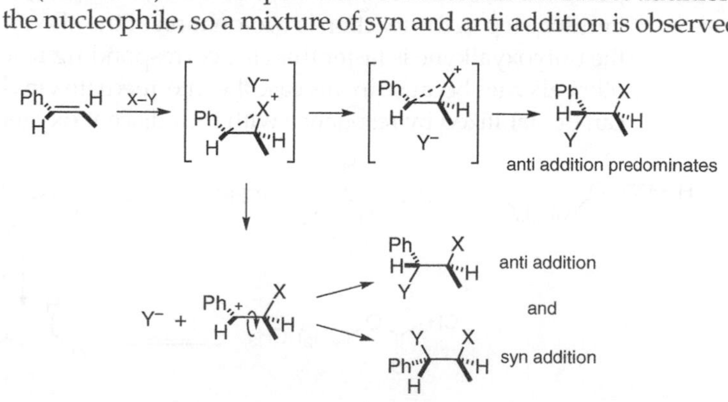

anti addition predominates

anti addition

and

syn addition

10.13. Rolston, J. H.; Yates, K. *J. Am. Chem. Soc.* **1969**, *91*, 1469. Styrene reacts through a benzylic carbocation or bridged cation with considerable benzylic carbocation character. The dimethylstyrene may give a bridged ion or a competition between benzyl and tertiary carbocations. Therefore, using *p*-nitro or other electron-withdrawing groups in the dimethyl isomer might shift the product distribution even more.

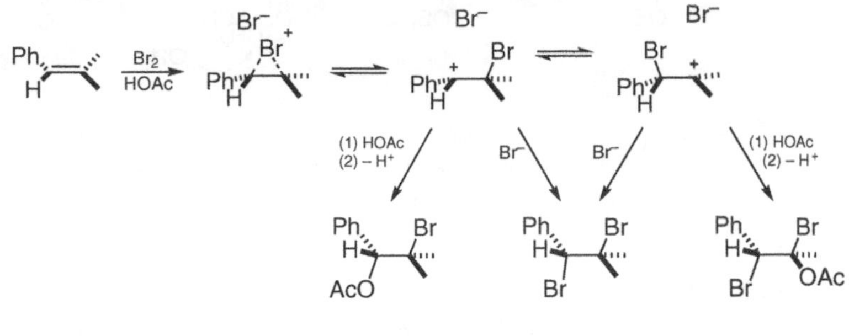

10.14. Cabaleiro, M. C.; Johnson, M. D. *J. Chem. Soc. B* **1967**, 565. The reaction proceeds through chlorocarbocation ion pairs **A** and **B** that can undergo conformational change before reaction with Cl^- to form III and IV by syn addition or reaction with solvent to form I and II by anti addition. In the presence of added chloride ion, anti addition of Cl^- is also observed.

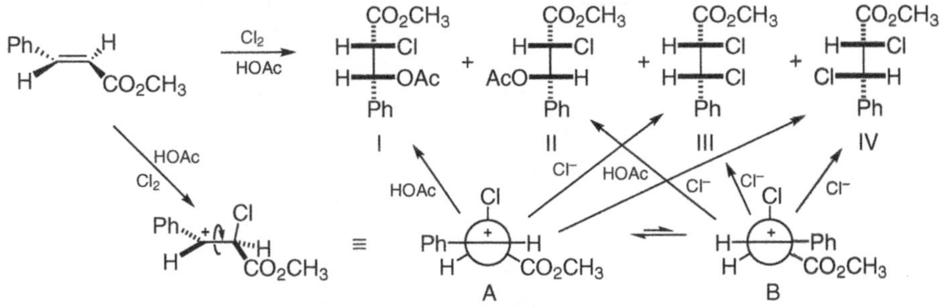

10.15. Rozen, S.; Brand, M. *J. Org. Chem.* **1986**, *51*, 3607. The product is *threo*-3,4-difluorohexan-1-ol acetate, formed by syn addition of F_2 to the alkene.

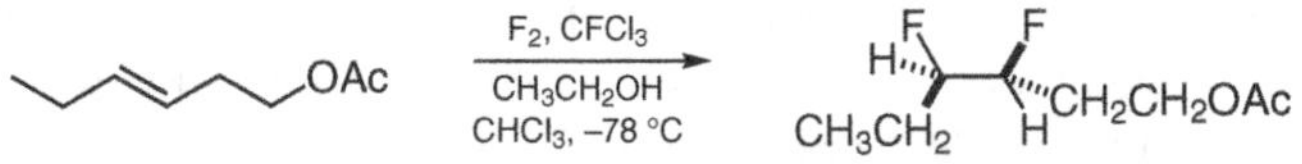

10.16. Jacobs, T. L.; Searles, S., Jr. *J. Am. Chem. Soc.* **1944**, *66*, 686. The rate-limiting step in the reaction is protonation of the alkyne. Reaction of the butyoxyalkyne is faster than the corresponding reaction of most other alkynes because in this case the intermediate vinyl carbocation can be stabilized by resonance with the adjacent oxygen.

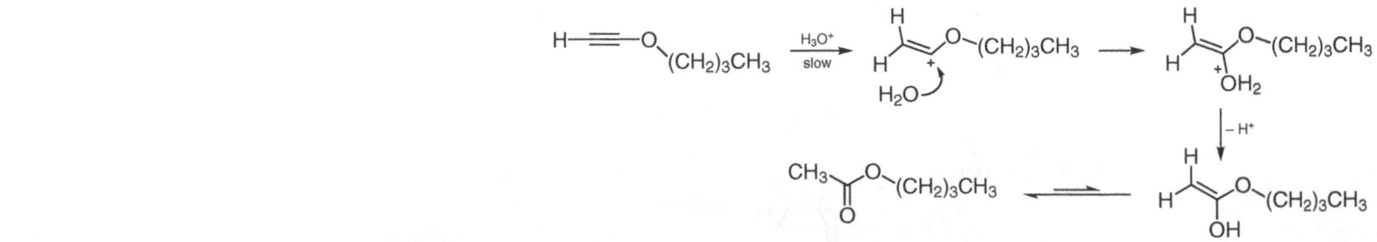

10.17. Halpern, J.; Tinker, H. B. *J. Am. Chem. Soc.* **1967**, *89*, 6427.

a. In general, the rate constants for oxymercuration correlate with Taft σ^* values. However, 1-penten-5-ol reacts by a pathway involving anchimeric assistance, so reactivity is enhanced. The products were isolated as the organomercury iodides.

b.

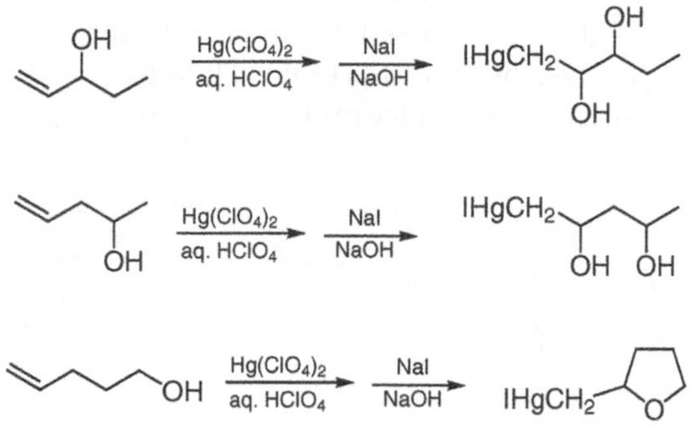

10.18. Kabalka, G. W.; Newton, R. J., Jr.; Jacobus, J. *J. Org. Chem.* **1978**, *43*, 1567. Syn addition leads to the formation of the threo product from the (*E*) alkene and the erythro product from the (*Z*) alkene.

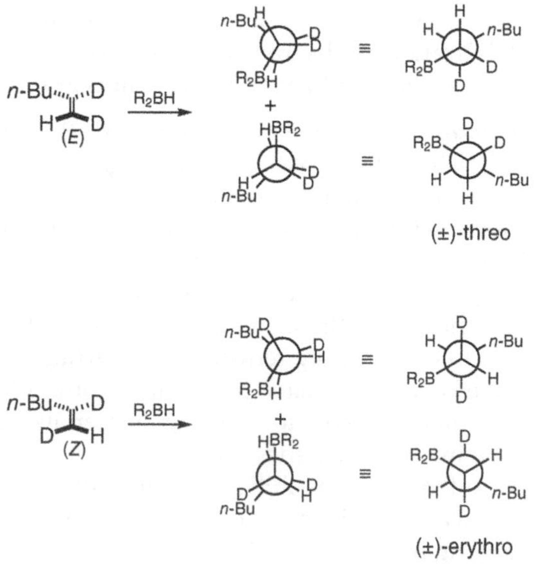

(±)-threo

(±)-erythro

10.19. a. Brown, H. C.; Geoghegan, P., Jr. *J. Am. Chem. Soc.* **1967**, *89*, 1522.

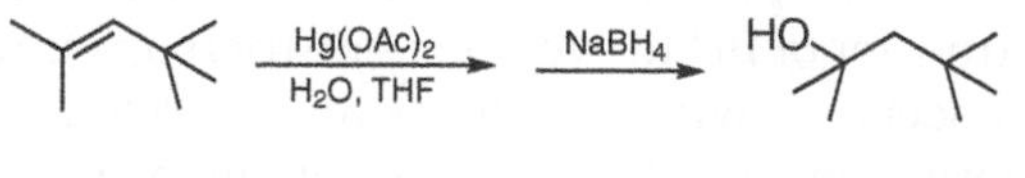

b. Brown, H. C.; Hammar, W. J. *J. Am. Chem. Soc.* **1967**, *89*, 1524.

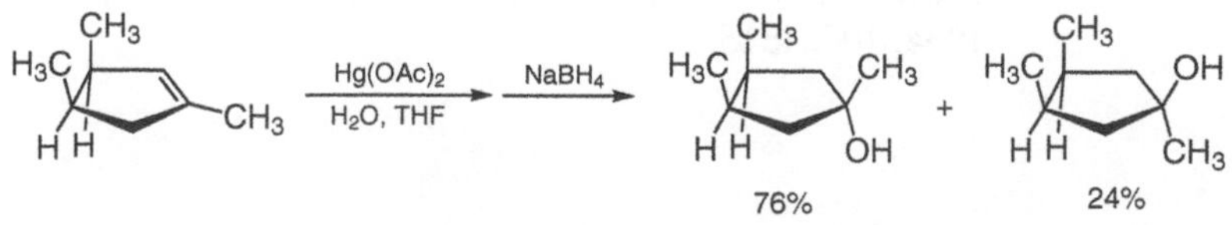

76%

24%

As indicated by the equation above, hydration occurs preferentially on the sterically less hindered face of the molecule.

10.20. Brown, H. C.; Kurek, J. T. *J. Am. Chem. Soc.* **1969,** *91,* 5647.

a. The product is *N*-cyclohexylacetamide.

b. The mechanism involves nucleophilic addition of CH_3CN to the mercurinium ion.

c. $Hg(NO_3)_2$ is used because nitrate is less nucleophilic than acetate or trifluoracetate. Therefore, it does not compete with solvent for nucleophilic addition to the mercurinium ion.

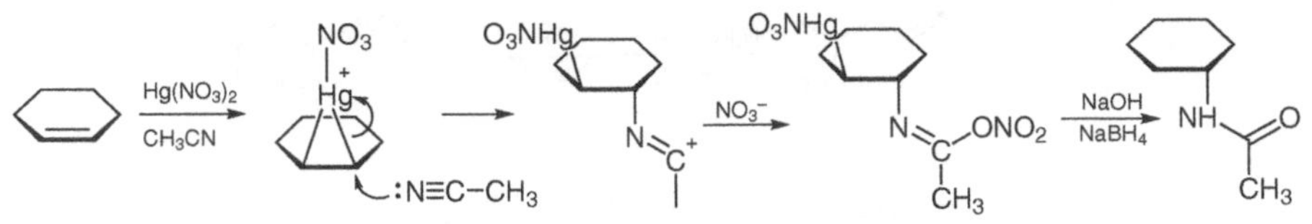

10.21. Swern, D. *J. Am. Chem. Soc.* **1947,** *69,* 1692. In each case, the double bond with the methyl substituent or without the electron-withdrawing group is more reactive.

10.22. Modro, A.; Schmid, G. H.; Yates, K. *J. Org. Chem.* **1977,** *42,* 3673. The formation of the intermediate bromonium ion–bromide ion pair is facilitated by a polar solvent. In a less polar solvent, stabilization by electron-donating alkyl groups is more significant because the stabilization by solvent is less effective.

10.23. Ruasse, M.-F.; Motallebi, S.; Galland, B. *J. Am. Chem. Soc.* **1991,** *113,* 3440. In both cases a neutral organic substrate is converted to a positively charged intermediate that adds a nucleophile to form the final product. Electron-releasing substituents accelerate both reactions. Both the departure of the leaving group in solvolysis and the departure of Br^- are facilitated by electrophilic solvent (or by Br_2 in nonpolar solvent). Just as a carbocation–carbanion pair may return to the starting material in solvolysis, there is evidence that a bromonium–bromide (or tribromide) ion pair may return to starting materials in the electrophilic bromination of alkenes. The authors of the paper cited here also considered the possibility of a range of electrophilic bromination pathways, analogous to the range of substitution pathways ranging from S_N1 to S_N2.

10.24. Bromine-assisted loss of brosylate in the reactant produces a cyclohexene bromonium ion that can react with bromide ion to produce racemic *trans*-1,2-dibromocyclohexane or with acetic acid to produce racemic *trans*-2-bromocyclohexyl acetate. Transfer of Br^+ to cyclopentene (perhaps through liberation of Br_2 associated with the reversal of cyclohexene bromonium ion formation) produces the cyclopentene bromonium ion and then the analogous 2-bromocyclopentyl bromide and acetate. See Brown, R. S.; Gedye, R.; Slebocka-Tilk, H.; et al. *J. Am. Chem. Soc.* **1984,** *106,* 4515.

10.25. The data suggest a radical chain reaction in which $Ph_2(O)P\cdot$ (formed by reaction of O_2 with $Ph_2(O)PH$) adds to the terminal carbon of the olefin, followed by reaction of the resulting 2° radical with $Ph_2(O)PH$ to form the "anti-Markovnikov" adduct and regenerate $Ph_2(O)P\cdot$. Addition of 4-*t*-butylcatechol inhibits the radical chain reaction and prevents product formation. See Hirai, T.; Han, L.-B. *Org. Lett.* **2007**, *9*, 53.

10.26. Dissociation of the (2,6-disubstituted pyridine)bromonium triflate gives an ion that can transfer Br^+ to the olefin. See Cui, X.-L.; Brown, R. S. *J. Org. Chem.* **2000**, *65*, 5653.

H3C H3C H3C N-Br-N CH3 H3C H3C ⁻OTf ⇌ H3C N CH3 CH3 + Br–N CH3 H3C H3C ⁻OTf

Br–N CH3 + OH → Br+ OH + N CH3

Br+ OH = [HO Br+]‡ → –H+ → Br O

10.27. Smith, W. B. *J. Org. Chem.* **1998**, *63*, 2661.

Br2 Br + Br⁻ Br + Br⁻ Br Br

10.28. Steinfeld, G.; Lozan, V.; Kersting, B. *Angew. Chem. Int. Ed.* **2003**, *42*, 2261. There is syn addition in the encapsulated complex. Apparently the initial π complex leading to the bromonium ion cannot form, so there is instead formation of a carbocation–bromide ion pair that collapses to the observed product.

10.29. Bach, R. D.; Glukhovtsev, M. N.; Canepa, C. *J. Am. Chem. Soc.* **1998**, *120*, 775.

= + O=N O H–O → [O=N O H O H H H H]‡ → O + H O–N O

10.30. The reaction is an Ad_EAr reaction. See Choi, H. Y.; Srisook, E.; Jang, K. S.; et al. *J. Org. Chem.* **2005**, *70*, 1222. The first step is an S_EAr reaction, resulting in a bromine substitution para to the methoxy group. Formation of the σ-complex by addition of Br^+ to the position ortho to methoxy gives a resonance-stabilized Wheland intermediate that cannot easily rearomatize, so attack by methanol (or methoxide) gives the addition product.

10.31. Bellucci, G.; Bianchini, R.; Chiappe, C.; et al. *J. Am. Chem. Soc.* **1988**, *110*, 546. Anchimeric assistance leads to a bromonium that reacts with bromide ion to give a dibromo compound with retention of stereochemistry. (Note that all structures are racemic.)

COCH3 N OH Br HBr CHCl3 COCH3 N Br Br

HBr

COCH3 N H_2O^+ Br + Br^- → COCH3 N Br^- Br^+

10.32. This was described as a stereospecific transannular cycloaddition by Gipson, R. M.; Guin, H. W.; Simonsen, S. H.; et al. *J. Am. Chem. Soc.* **1966**, *88*, 5366. Mechanism (a) assumes formation of a bromonium ion before cyclization. Mechanism (b), from the reference cited here, shows the alternative single-step process:

(a) $HOCH_3$ R R R R H H H H Br^+ (b) $HOCH_3$ R R R R H H H H Br Br

R = CO_2Et

10.33. Bellucci, G.; Bianchini, R.; Chiappe, C.; et al. *J. Am. Chem. Soc.* **1995**, *117*, 6243. Steric hindrance to nucleophilic attack of bromide on the bromonium ion leads to proton removal and alkene formation.

t-Bu t-Bu Br_2 CD_2Cl_2 → t-Bu H Br_3^- Br^+ t-Bu → t-Bu Br t-Bu

10.34. Bojase, G.; Nguyen, T. V.; Payne, A. D.; et al. *Chem. Sci.* **2011**, *2*, 229. The authors propose a one-step reaction proceeding through a transition state in which the cyclopropane rings react in a cooperative manner.

[H^+ $^-O_2CCF_3$]‡

10.35. Jin, T.; Himuro, M.; Yamamoto, Y. *Angew. Chem. Int. Ed.* **2009**, *48*, 5893.

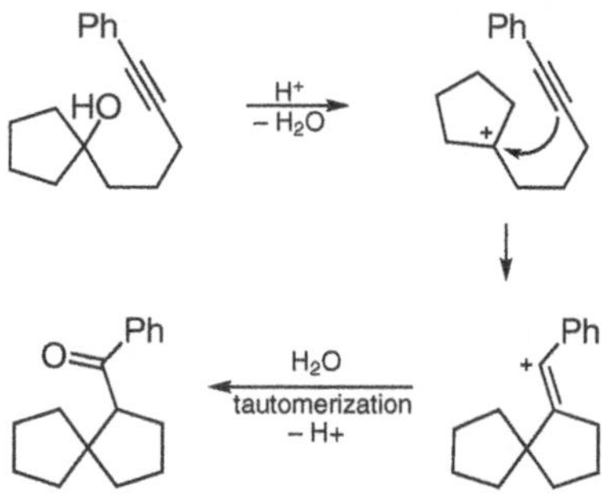

10.36. It may be argued that—for some purposes, at least—overly simplified mechanistic models allow synthetic chemists to focus on synthetic strategy and not on detailed representations of reactive intermediates. Ultimately, however, representations of mechanisms and reactive intermediates corresponding to models developed through mechanistic studies may be needed to design novel synthetic reactions.

10.37. Markovnikov's rule is a useful tool to organize patterns of chemical reactivity, but it must be presented as a simple model and not as a "rule" to be memorized. The same is true of "rules" associated with others, such as Bredt, Saytzeff, and Hofmann.

CHAPTER 11

Pericyclic Reactions

11.1. For the thermal reaction, the HOMO of octatetraene is ψ_4. There are three nodes in ψ_4, so the sign of the coefficient for ϕ_1 is opposite that of ϕ_8. Therefore, the thermal reaction should be conrotatory. In the excited state, HOMO is ψ_5. Now the sign of the coefficient for ϕ_1 is the same as that for ϕ_8, so the photochemical reaction should be disrotatory.

11.2. The thermal [1,7] hydrogen shift is analyzed in terms of a transition structure resembling a hydrogen atom and a heptatrienyl radical. For the heptatrienyl radical, HOMO is ψ_4, which has three nodes. The coefficient for ϕ_1 has the opposite sign from that of ϕ_7. Therefore, in order to maintain a bonding relationship from H to C1 while a bond from H to C7 is forming, the reaction would have to be antarafacial with respect to the heptatrienyl radical.

The thermal [1,9] hydrogen shift is analyzed in terms of a transition structure resembling a hydrogen atom and a nonatetraenyl radical. For the nonatetraenyl radical, HOMO is ψ_5. There are four nodes in ψ_5, so the coefficients of ϕ_1 and ϕ_9 have the same sign. Therefore, a bonding relationship between H and C1 can be maintained while a bonding relationship between H and C9 is being established, so the suprafacial [1,9] hydrogen shift is allowed by the principles of orbital symmetry.

11.3. The antarafacial–antarafacial transition structure is shown below:

antarafacial-
antarafacial

The problem is analyzed by considering ψ_2 of one allyl fragment (drawn here at the back of the transition structure) and ψ_3 of the other allyl fragment (drawn here at the front of the transition structure). It may be seen that an antibonding relationship exists

Solutions Manual for Perspectives on Structure and Mechanism in Organic Chemistry, Third Edition.
Felix A. Carroll.

between the orbitals of the incipient C1–C6 bond, so the reaction is forbidden by the principles of orbital symmetry.

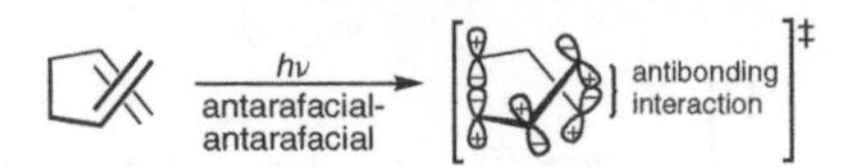

11.4. The suprafacial–suprafacial [$_{\pi}2_{s}$ + $_{\pi}6_{s}$] cycloaddition is analyzed in terms of a transition structure with a plane of symmetry (bisecting the C3–C4 bond of hexatriene and the C1−C2 bond of ethene) and is forbidden by the principles of orbital symmetry.

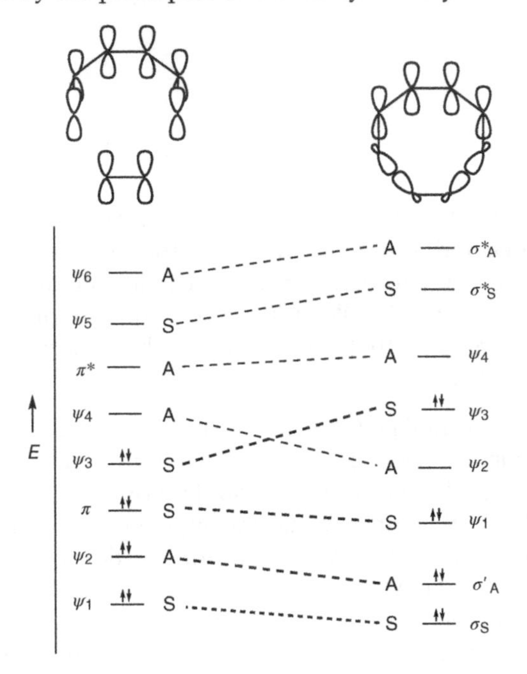

11.5. The suprafacial–suprafacial [$_{\pi}4_{s}$ + $_{\pi}4_{s}$] cycloaddition is analyzed in terms of a transition structure with a plane of symmetry (bisecting the C2–C3 bond in each butadiene molecule) and is forbidden by the principles of orbital symmetry.

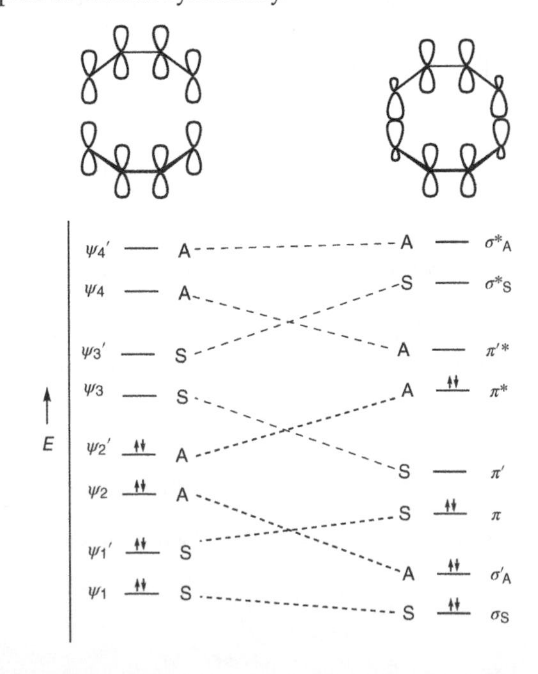

11.6. Longuet-Higgins, H. C.; Abrahamson, E. W. *J. Am. Chem. Soc.* **1965**, *87*, 2045. A discussion of the construction of orbital correlation diagrams using HMO theory was given by Dalton, J. C.; Friedrich, L. E. *J. Chem. Educ.* **1975**, *52*, 721. For the cyclopropyl cation, only σ is populated. A plane of symmetry is maintained in the disrotatory reaction, so a symmetric σ orbital correlates with ψ_1 of the allyl cation and the reaction is allowed. For the conrotatory pathway, a C_2 rotation axis is maintained. Now the σ orbital is symmetric but ψ_1 of allyl is antisymmetric. Therefore, the σ orbital correlates with ψ_2 of allyl, so the reaction is thermally forbidden.

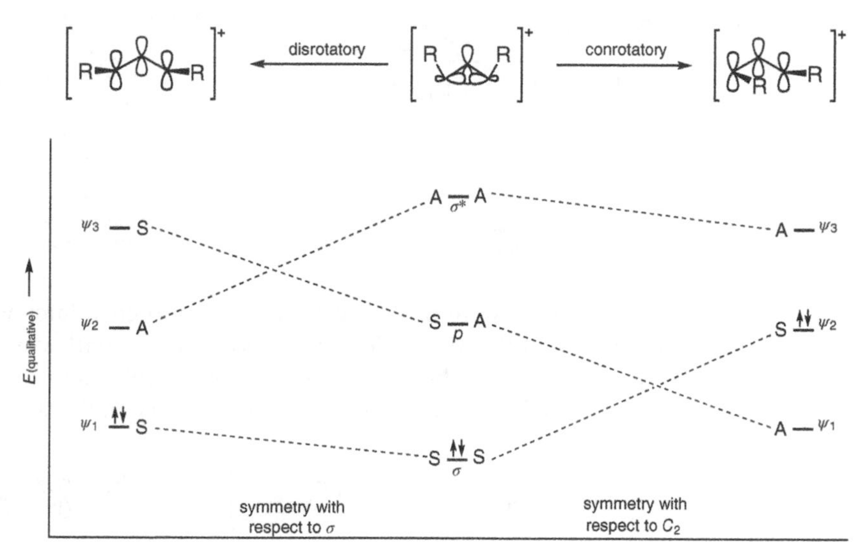

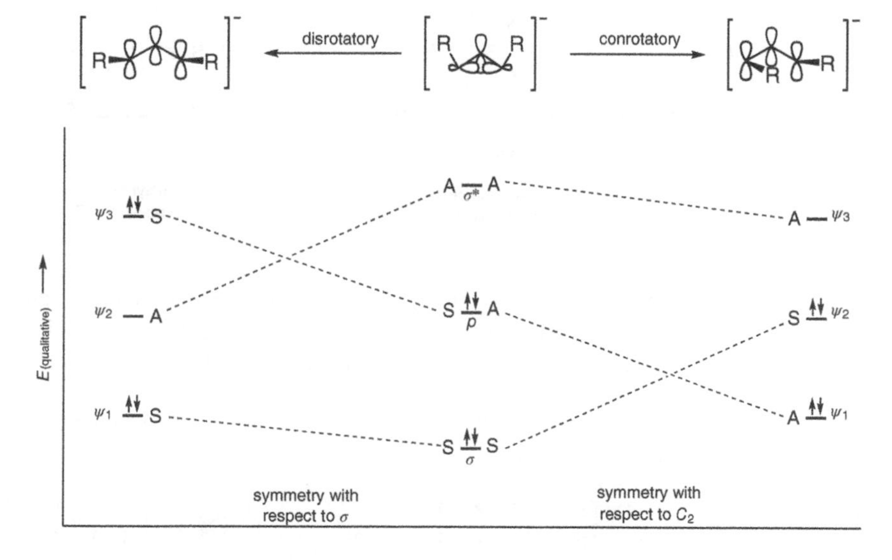

In the anion, both σ and p of the cyclopropyl moiety are populated, and their orbital symmetries are designated as shown above. Now the cyclopropyl anion correlates with the allyl anion by the conrotatory pathway.

The state correlation diagrams are constructed from the molecular orbital correlation diagrams. For the cation, it is evident that the ground state (GS) of the cyclopropyl cation correlates with the ground state of the allyl cation for the disrotatory opening. (The state symmetry designations are the products of the MO symmetry designations for the molecular orbitals populated in each state.) However, the cyclopropyl cation correlates with the allyl cation for a photochemical reaction because ES-1 of cyclopropyl correlates with ES-1 of allyl for that process.

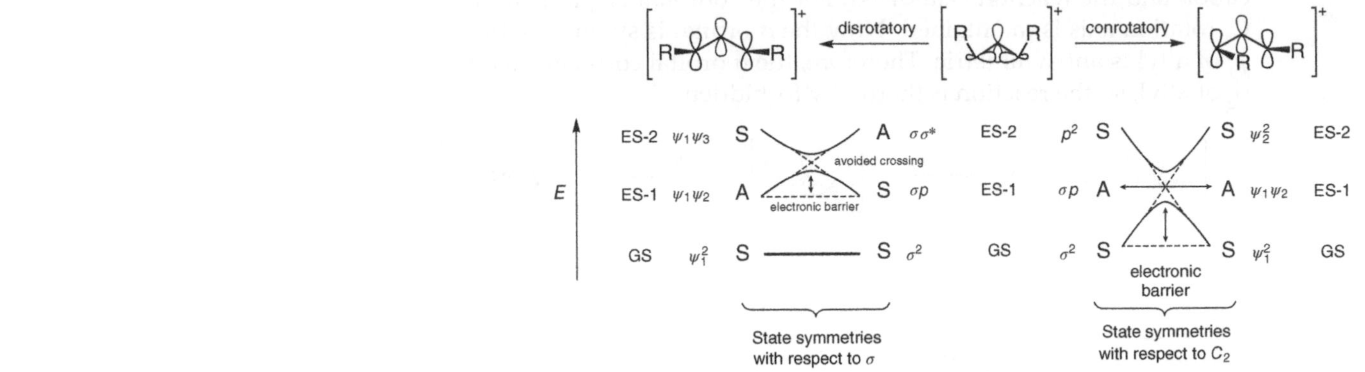

The ground state of the cyclopropyl anion correlates with the ground state of the allyl anion by the conrotatory pathway, but the first excited state of the cyclopropyl anion correlates with the first excited state of the allyl anion by the disrotatory pathway.

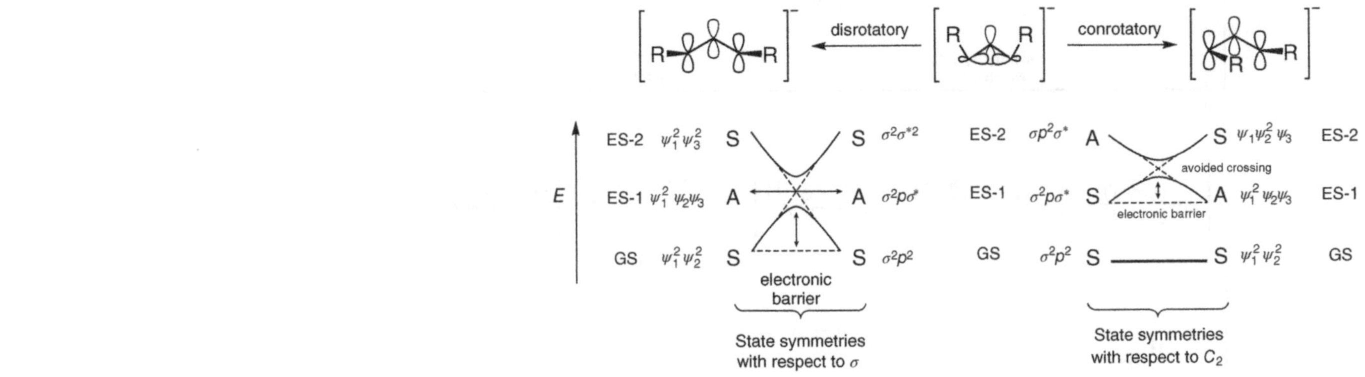

11.7. a. The Hückel MO energy levels for cyclopropenyl are calculated on the left below. The Möbius MO energy levels for cyclopropenyl are calculated on the right.

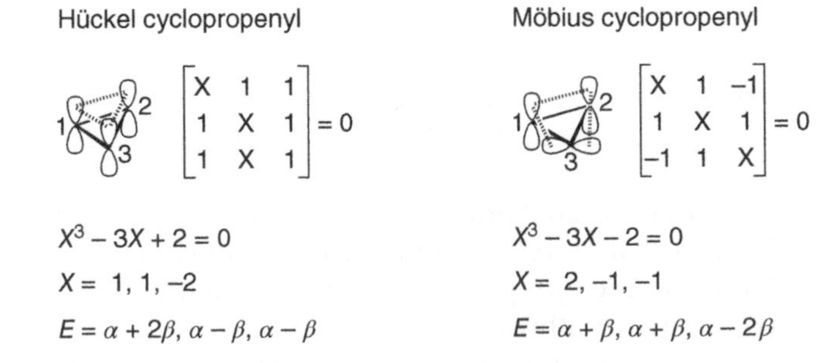

b. For the cation:

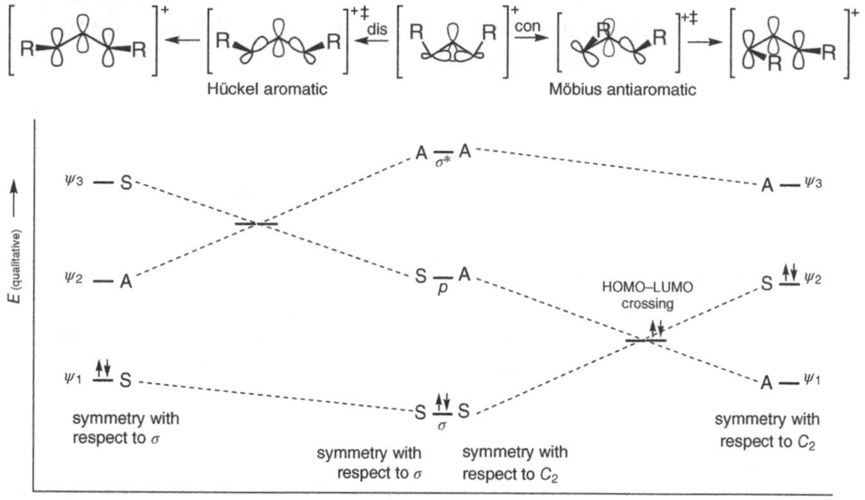

For the anion:

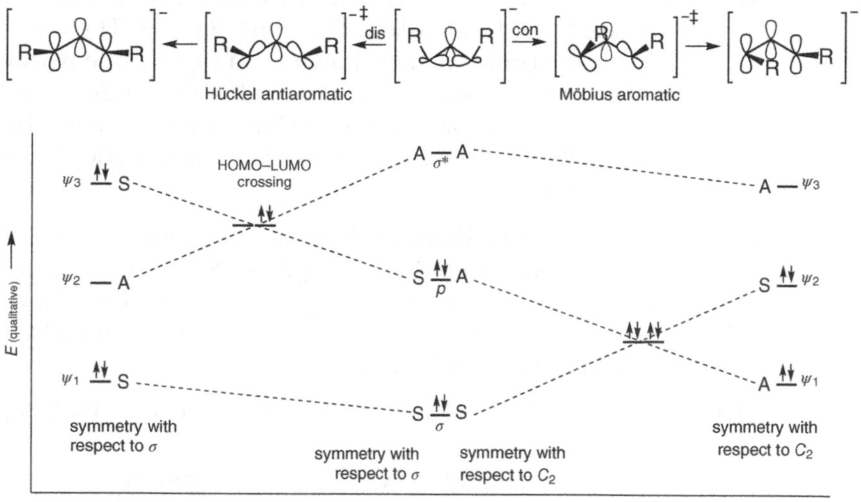

11.8. **a.** The pathway shown is a $[_{\pi}4_{a}]$ process. The number of $(4n+2)_{s}$ components is 0, the number of $(4r)_{a}$ components is 1, the total is odd, and the reaction is allowed. The transition structure should be Möbius aromatic.

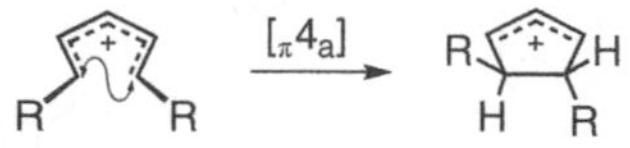

b. The pathway shown is a $[_{\pi}6_{s}]$ process. The number of $(4n+2)_{s}$ components is 1, the number of $(4r)_{a}$ components is 0, the total is odd, and the reaction is allowed. The transition structure should be Hückel aromatic.

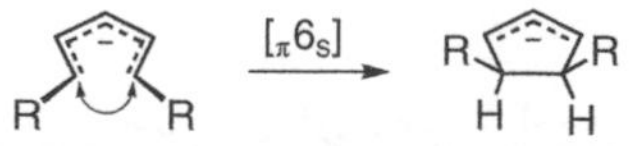

c. The pathway shown is a $[_{\pi}6_s + _{\pi}4_s]$ process. The number of $(4n+2)_s$ components is 1, the number of $(4r)_a$ components is 0, the total is odd, and the reaction is allowed. The transition structure should be Hückel aromatic.

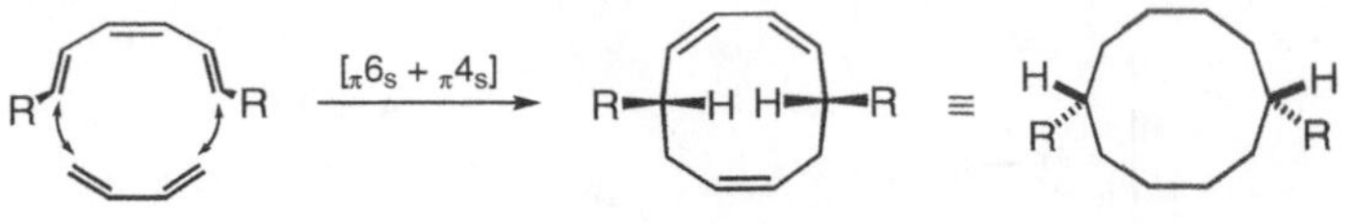

11.9. Fráter, G.; Schmid, H. *Helv. Chim. Acta* **1968**, *51*, 190. The authors categorized the reaction as a [5,5]-sigmatropic rearrangement, which is an allowed suprafacial–suprafacial process, followed by tautomerization of the cyclohexadienone to the phenol.

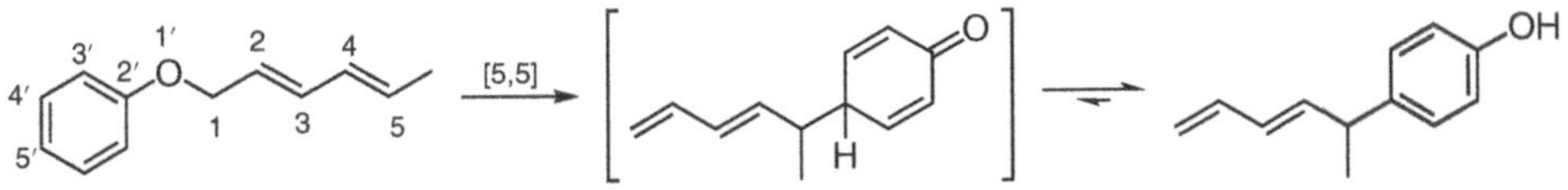

11.10. Vogel, E. *Liebigs Ann. Chem.* **1958**, *615*, 1; Also see Hammond, G. S.; DeBoer, C. D. *J. Am. Chem. Soc.* **1964**, *86*, 899. The cis isomer can achieve the conformation shown for a rapid Cope rearrangement. In the trans isomer, the two trans double bonds cannot become aligned properly for a concerted Cope rearrangement. Therefore, it is likely that the trans isomer reacts by homolytic cleavage of the cyclobutane σ bond.

11.11. Doering, W. v. E.; Wiley, D. W. *Tetrahedron* **1960**, *11*, 183. The reaction shown in equation 11.57 is a $[_{\pi}2_s + _{\pi}8_s]$ cycloaddition, which is allowed by a suprafacial–suprafacial pathway. However, the reaction in equation 11.58 is a $[_{\pi}2_s + _{\pi}6_s]$ cycloaddition, which requires an SA or AS pathway.

11.12. **a.** Bates, R. B.; McCombs, D. A. *Tetrahedron. Lett.* **1969**, *10*, 977.

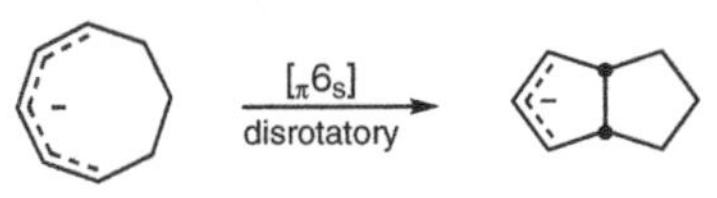

b. Pomerantz, M.; Wilke, R. N.; Gruber, G. W.; Roy, U. *J. Am. Chem. Soc.* **1972**, *94*, 2752.

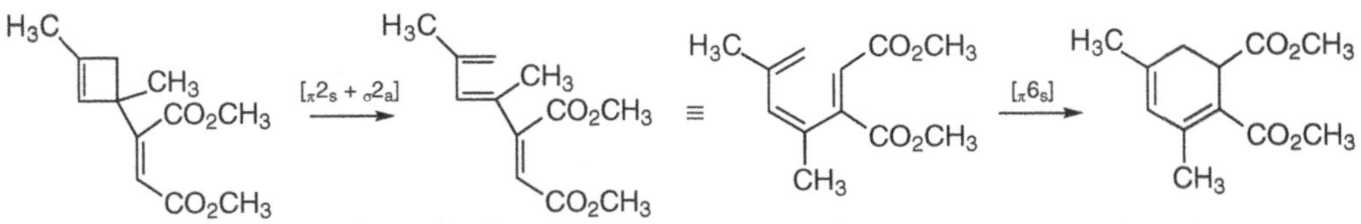

c. Arnold, B. J.; Sammes, P. G. *J. Chem. Soc. Chem. Commun.* **1972**, 1034; Arnold, B. J.; Sammes, P. G.; Wallace, T. W. *J. Chem. Soc. Perkin Trans. 1* **1974**, 415.

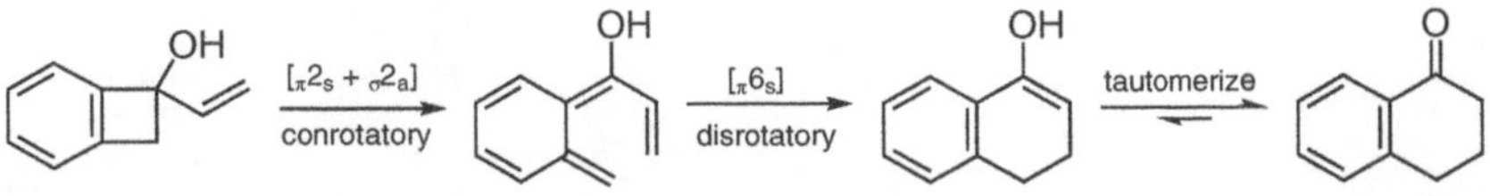

11.13. Brown, J. M. *Chem. Commun.* **1965**, 226. The Wittig reaction produces *cis*-6-vinylbicyclo[3.1.0]hex-2-ene, which undergoes a Cope rearrangement to give the product.

H H $Ph_3\overset{+}{P}CH_3\overset{-}{X}$ *t*-BuOK DMSO H_2C H H H [3,3]

11.14. Curtin, D. Y.; Johnson, H. W., Jr. *J. Am. Chem. Soc.* **1956**, *78*, 2611. The intermediate cyclohexadienone can undergo a retro-Claisen rearrangement with migration of the methallyl group or can undergo a Cope rearrangement with either the allyl or methylallyl group.

[3,3] O [3,3] [3,3] [3,3] O O O H H tautomerization tautomerization HO HO

11.15. Dowd, P.; personal communication to Woodward and Hoffmann (Woodward, R. B.; Hoffmann, R. *The Conservation of Orbital Symmetry*; Verlag Chemie/Academic Press: Weinheim, 1971; p. 143). Also, see Fleming, I.; personal communication to Woodward and Hoffmann (*ibid.*). The reaction is allowed, with $m = 4$ and $n = 2$.

11.16. Arnett, E. M. *J. Org. Chem.* **1960**, *25*, 324. The product is formed by an ene reaction between benzyne and the diene.

H H H ‡ Ph

11.17. Copley, S. D.; Knowles, J. R. *J. Am. Chem. Soc.* **1985**, *107*, 5306. Also see Dagani, R. *Chem. Eng. News* 1984 (May 28), 26. For a more recent study of the chorismate to prephenate reaction, see Claeyssens, F.; Ranaghan, K. E.; Lawan, N.; et al. *Org. Biomol. Chem.* **2011**, *9*, 1578.

a. The rearrangement is a [3,3]-sigmatropic rearrangement:

[3,3]

b. The diastereomers are shown here, where "T" represents ^3H.

(E)
(T)

[3,3]
(boat)

[3,3]
(chair)

(E)
[3,3]
(chair)
(R)

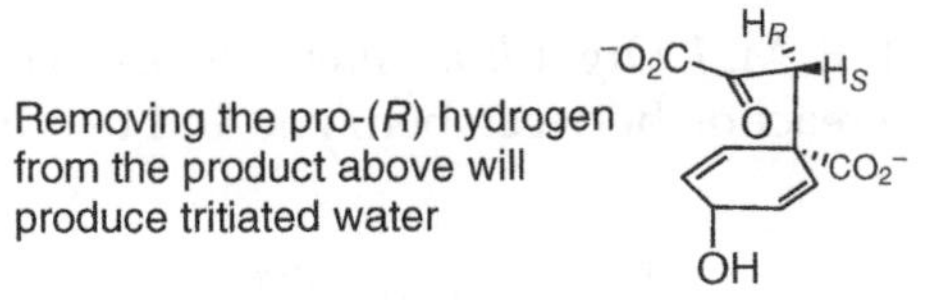
Removing the pro-(R) hydrogen
from the product above will
produce tritiated water

11.18. Conroy, H.; Firestone, R. A. *J. Am. Chem. Soc.* **1956**, *78*, 2290. Products having molecular formula $C_{15}H_{16}O_4$ are consistent with (racemic) Diels–Alder adducts resulting from reaction of maleic anhydride and 6-allyl-2,6-dimethyl-2,4-cyclohexadieneone. Their formation suggests that the cyclohexadieneone is an intermediate in the thermal conversion of allyl 2,6-dimethylphenyl ether to 4-allyl-2,6-dimethylphenol by consecutive [3,3]-sigmatropic rearrangements.

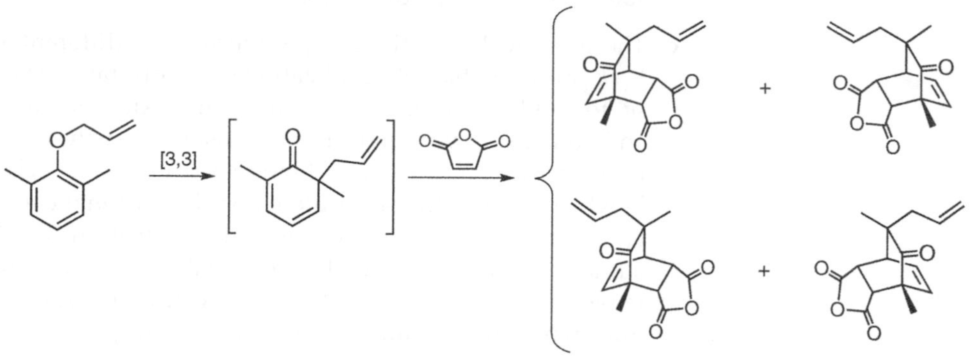

[3,3]

11.19. Heimgartner, H.; Hansen, H.-J.; Schmid, H. *Helv. Chim. Acta* **1972**, *55*, 1385. The first is a disrotatory electrocyclic closure, followed by a [1,5] hydrogen shift.

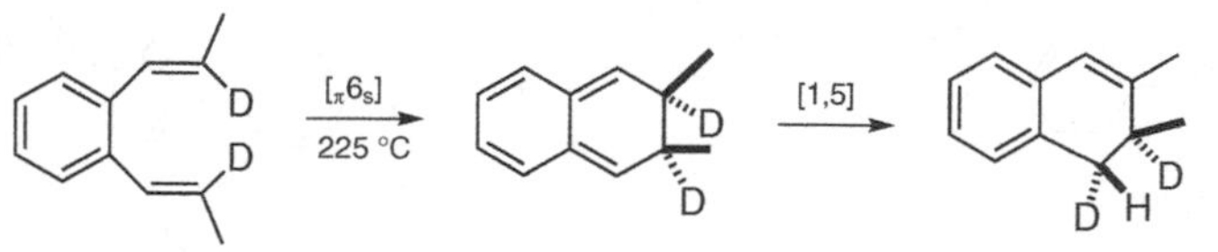

$[_\pi 6_s]$ 225 °C [1,5]

11.20. Greenwood, F. L. *J. Org. Chem.* **1959**, *24*, 1735. See also the discussion in DePuy, C. H.; King, R. W. *Chem. Rev.* **1960**, *60*, 431. The reactants equilibrate by a concerted reaction before elimination of acetic acid occurs.

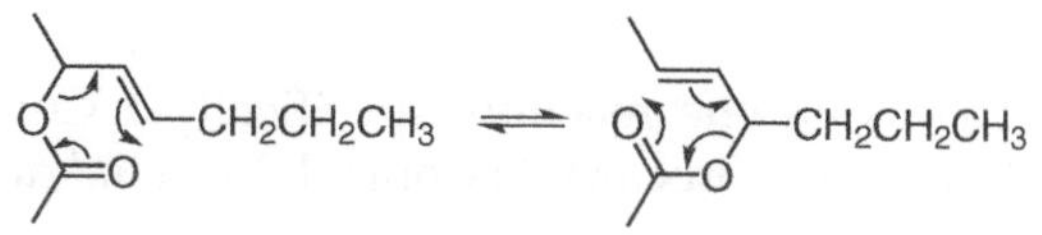

$-CH_2CH_2CH_3$ ⇌ $-CH_2CH_2CH_3$

11.21. Agosta, W. C. *J. Am. Chem. Soc.* **1964**, *86*, 2638. The product was formed by a Diels–Alder (suprafacial–suprafacial) reaction, so the configuration of the (−)-pentadienedioic acid must have been (*R*).

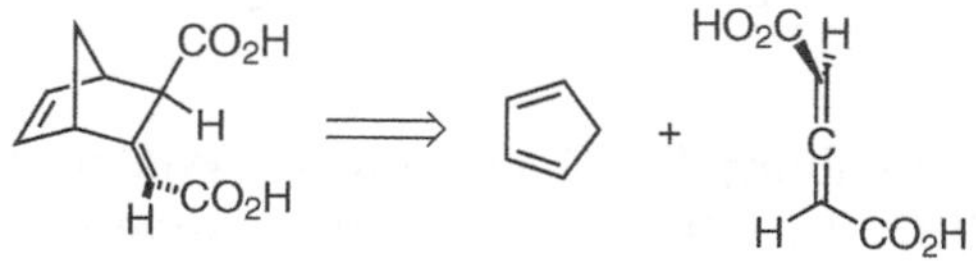

CO_2H H H CO_2H ⟹ + HO_2C H C H CO_2H

11.22. Baldwin, J. E.; Reddy, V. P.; Schaad, L. J.; et al. *J. Am. Chem. Soc.* **1988**, *110*, 8555.

a. The rate constant for formation of **135** is $1.1/(1.1+1.0)\times 8.17\times 10^{-5}$ $s^{-1} = 4.28\times 10^{-5}$ s^{-1}. Similarly, the rate constant for formation of **136** is $1.0/(1.1+1.0)\times 8.17\times 10^{-5}$ $s^{-1} = 3.89\times 10^{-5}$ s^{-1}.

b. $k_H/k_D = 4.465/4.28 = 1.04$ for formation of **135**. For formation of **136**, $k_H/k_D = 4.465/3.89 = 1.15$.

c. These are α 2° kinetic isotope effects. The different values of k_H/k_D indicate that rehybridization of the orbitals (formally from sp^3 to sp^2) has not occurred to the same extent for the "inside" and "outside" hydrogens in the transition structure. This result is inconsistent with the usual model for an electrocyclic reaction in which both of the substituents on the terminal carbon atoms have rotated to the same extent in the transition structure. The authors of the paper cited here calculated that the orbital to the inner hydrogen atom has 74% *p* character in the transition structure, while the orbital to the outer hydrogen atom has 68% *p* character.

11.23. **a.** [9,9]-sigmatropic shift. See Park, K. H.; Kang, J. S. *J. Org. Chem.* **1997**, *62*, 3794.

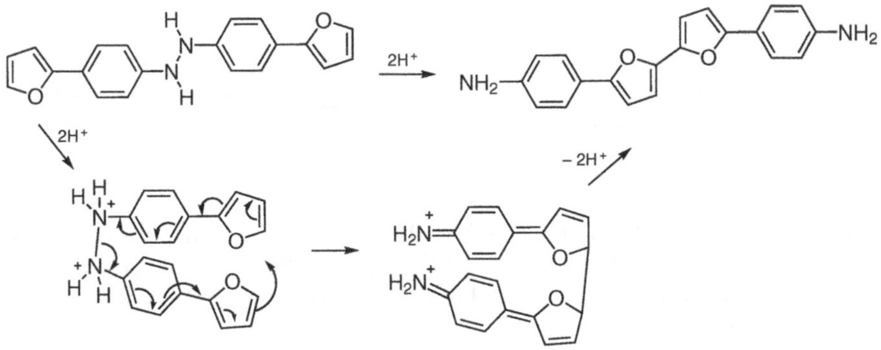

b. This is an ene reaction, specifically a Conia-Ene reaction. See Moinet, G.; Brocard, J.; Conia, J. M. *Tetrahedron Lett.* **1972**, 4461.

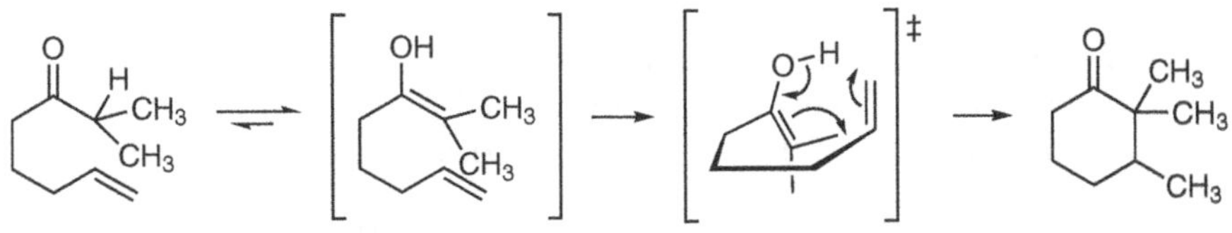

c. Schomburg, D.; Thielmann, M.; Winterfeldt, E. *Tetrahedron Lett.* **1985**, *26*, 1705. The reaction is proposed to occur in two steps. A Diels–Alder reaction is followed by a retro-Diels–Alder.

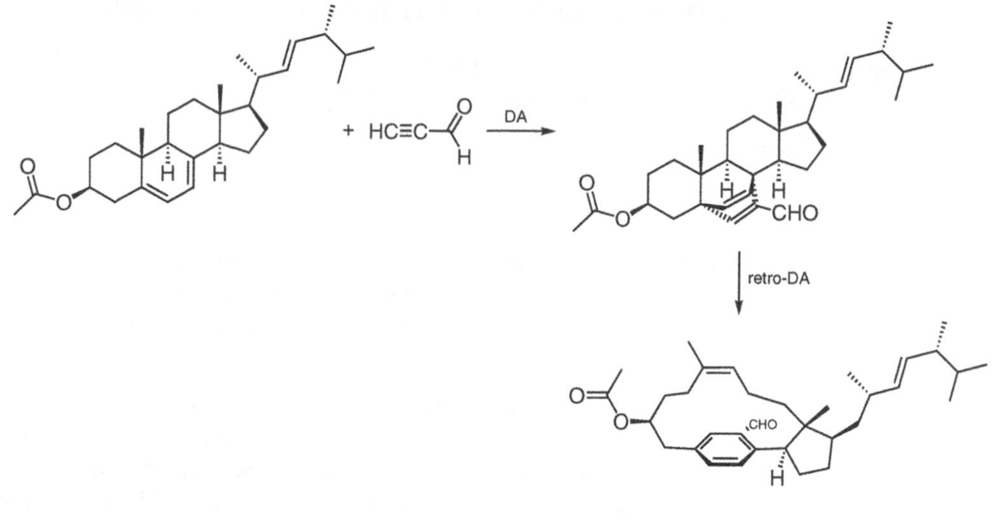

H H + HC≡C– O H DA → H H CHO

retro-DA

O O CHO H

d. The reaction was proposed to occur by a tandem Diels–Alder—ene synthesis via the intermediate shown here by Kraus, G. A.; Kim, J. *Org. Lett.* **2004**, *6*, 3115.

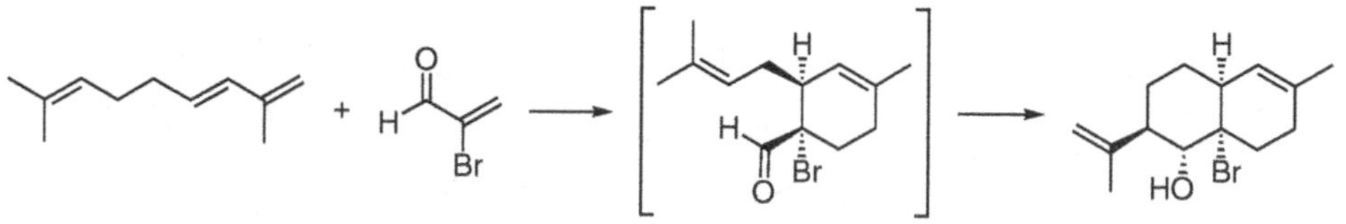

+ H O Br → [H H O Br] → H HO Br

e. See Ichikawa, Y.; Yamaoka, T.; Nakano, K.; et al. *Org. Lett.* **2007**, *9*, 2989. [3,3]-sigmatropic rearrangement.

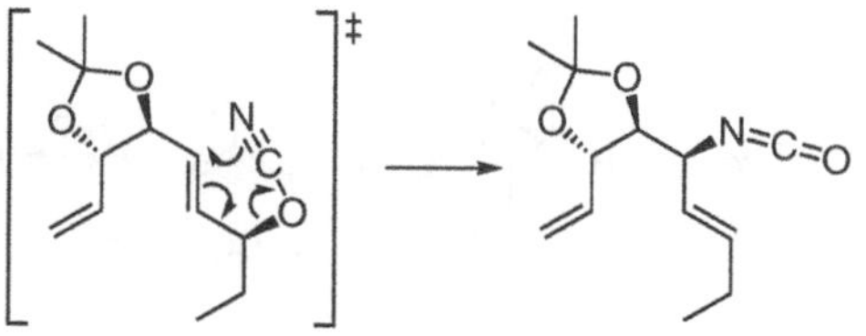

[O O N C O]‡ → O O N=C=O

f. Elimination of HCl from the reactant produces a chiral 2-H azirine, which subsequently undergoes a Diels–Alder reaction with cyclopentadiene. See Davis, F. A.; Deng, J. *Org. Lett.* **2007**, *9*, 1707.

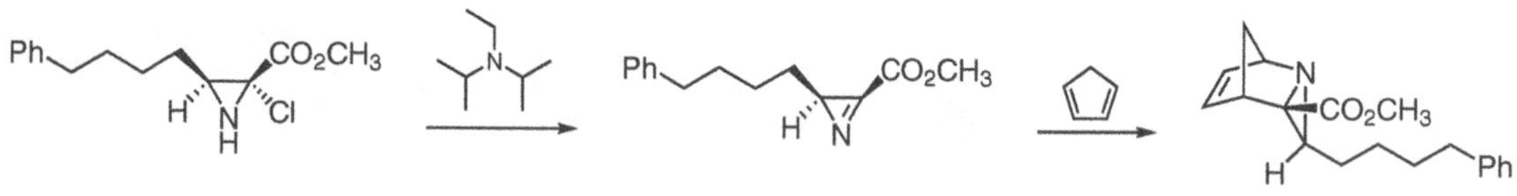

Ph CO_2CH_3 H N H Cl → N → Ph CO_2CH_3 H N → N CO_2CH_3 H Ph

g. Migawa, M. T.; Townsend, L. B. *Org. Lett.* **1999**, *1*, 537. Tautomerization is followed by a retro-Diels–Alder reaction.

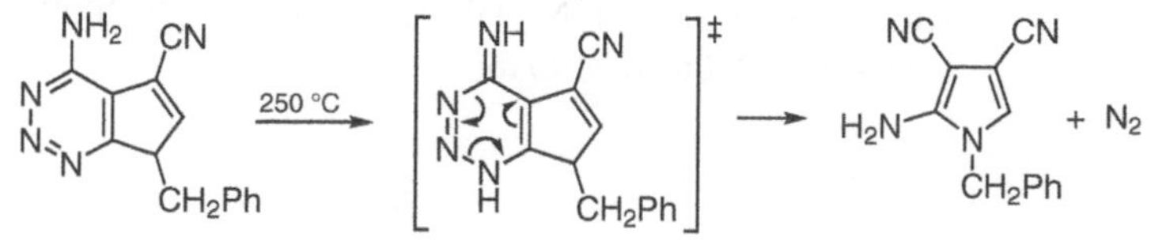

NH_2 CN N N N CH_2Ph 250 °C → [NH CN N N N H CH_2Ph]‡ → NC CN H_2N N CH_2Ph + N_2

11.24. Makosza, M.; Podraza, R. *Eur. J. Org. Chem.* **2000**, 193.

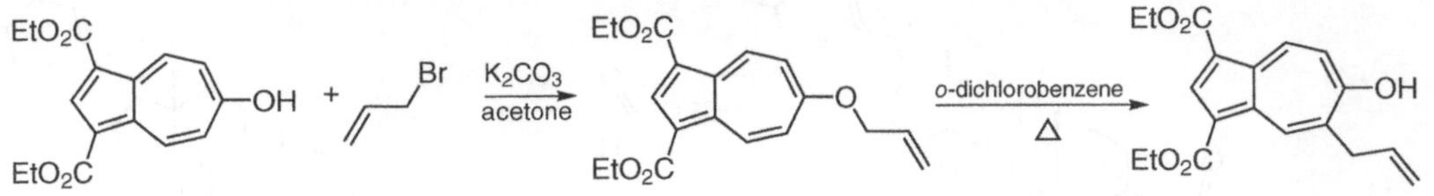

11.25. The suprafacial hydrogen shift is sterically impossible; alternative bimolecular pathways for rearrangement were proposed by Norton, J. E.; Northrop, B. H.; Nuckolls, C.; et al. *Org. Lett.* **2006**, *8*, 4915.

11.26. **a.** See Castro, C.; Karney, W. L.; Noey, E.; et al. *Org. Lett.* **2008**, *10*, 1287. Electrocyclic opening of the cyclobutene rings leads to the cyclohexenofused [12]annulene shown. A "Möbius π-bond shift" results in isomerization to a structure with one cis and two trans double bonds. A $[_{\pi}6_{s}]$ electrocyclic reaction produces the precursor, which undergoes intramolecular Diels–Alder reaction to give the final product.

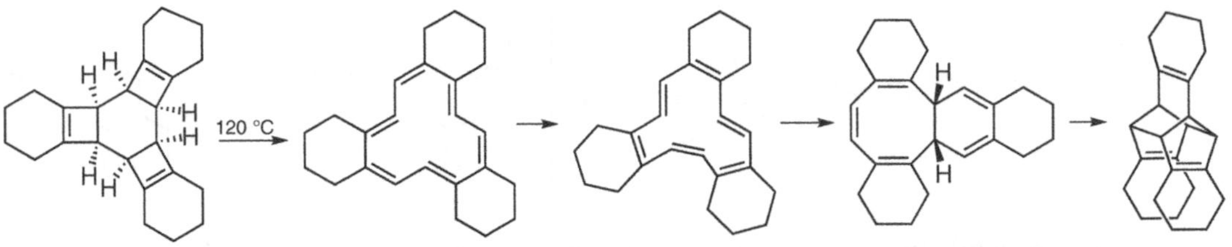

b. Carroll, M. F. *J. Chem. Soc.* **1940**, 704.

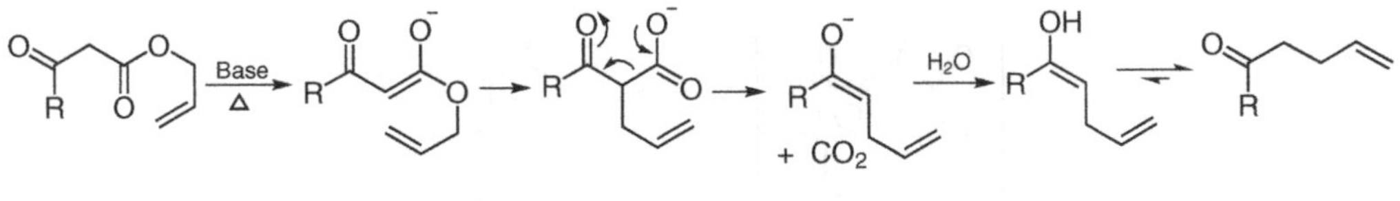

c. Proton removal by KH produces an anion that undergoes oxyanion-assisted retro-Diels–Alder to give an enolate that tautomerizes to the aldehyde. See Miyashi, T.; Ahmed, A.; Mukai, T. *J. Chem. Soc. Chem. Commun.* **1984**, 179.

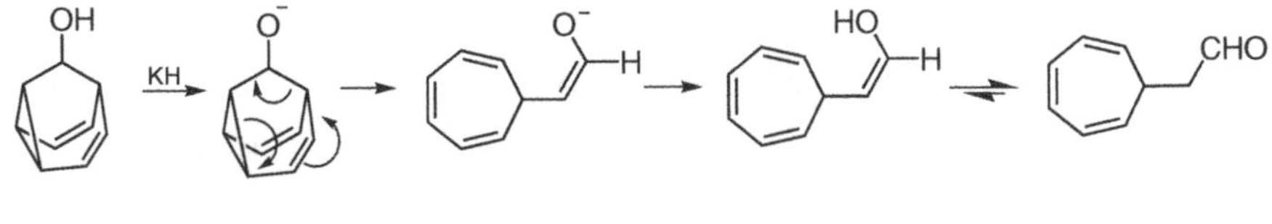

d. Silver-assisted solvolysis of the reactant produces a pentadienyl cation that closes to an allylic cyclopentenyl cation. Subsequent Friedel-Crafts reaction and further steps lead to the product shown. See Grant, T. N.; West, F. G. *Org. Lett.* **2007**, *9*, 3789; *J. Am. Chem. Soc.* **2006**, *128*, 9348.

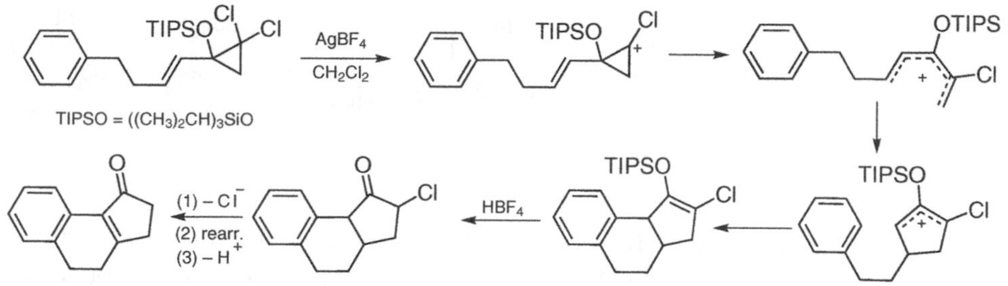

TIPSO = $((CH_3)_2CH)_3SiO$

11.27. Takao, K.; Munakata, R.; Tadano, K. *Chem. Rev.* **2005**, *105*, 4779.

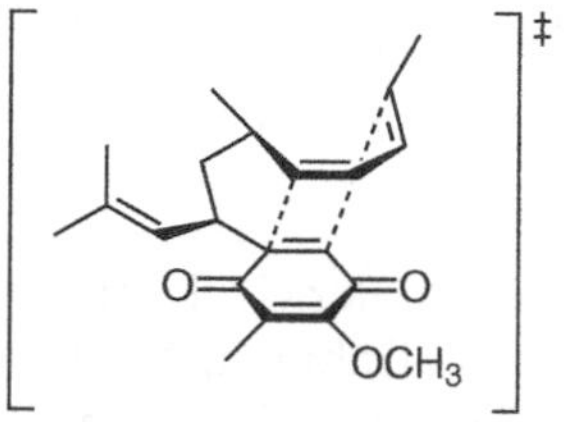

11.28. a. Dockendorff, C.; Sahli, S.; Olsen, M.; et al. *J. Am. Chem. Soc.* **2005**, *127*, 15028.

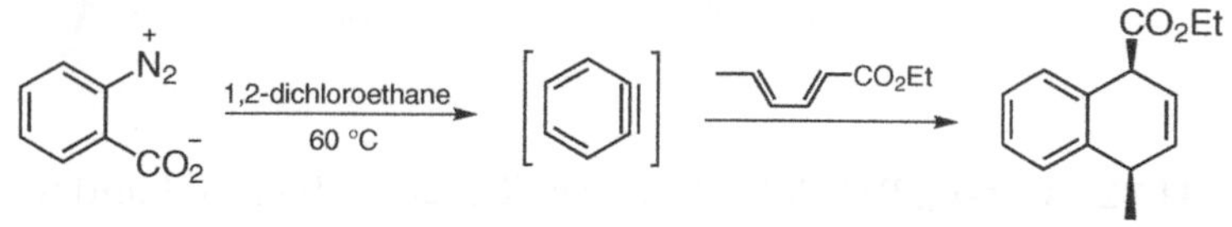

b. Intramolecular Diels–Alder. Jenkins, P. R. *J. Braz. Chem. Soc.* **1996**, *7*, 343.

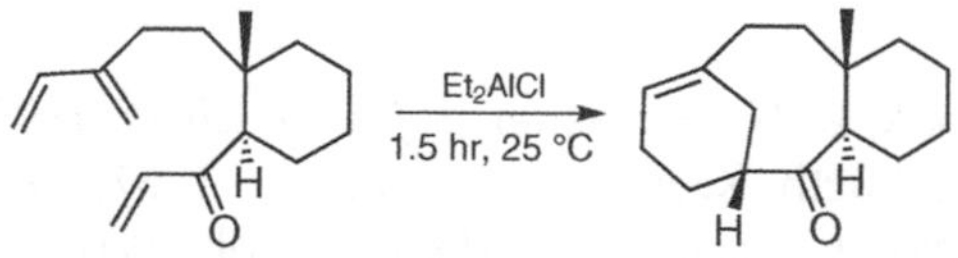

c. Cope rearrangement. Marvell, E. N.; Tao, T. *Tetrahedron Lett.* **1969**, 1341.

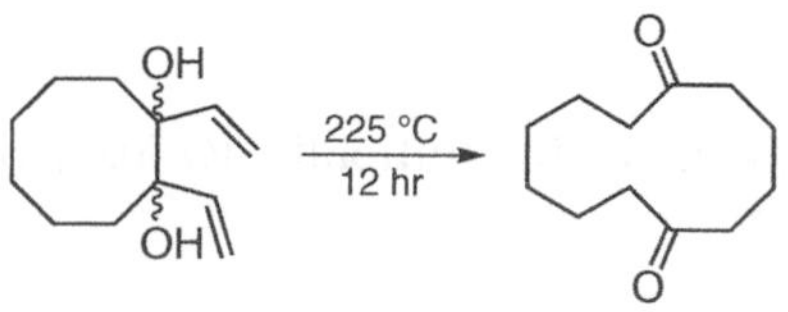

11.29. Williams, D. R.; Reeves, J. T.; Nag, P. P.; et al. *J. Am. Chem. Soc.* **2006**, *128*, 12339. The key step is the six-electron disrotatory cyclization of the pentadienyl anion to give the *cis*-fused bicyclo[3.3]octadienyl anion, which subsequently reacts with benzophenone to yield the final product.

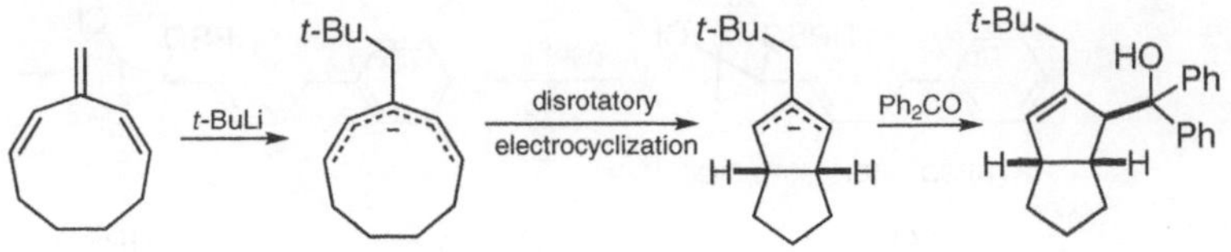

t-Bu t-BuLi disrotatory electrocyclization t-Bu H H Ph$_2$CO t-Bu HO Ph Ph H H

11.30. Lida, K.; Komada, K.; Saito, M.; et al. *J. Org. Chem.* **1999**, *64*, 7407. The cyclobutene moiety undergoes electrocyclic opening followed by a [1,5]-deuterium shift.

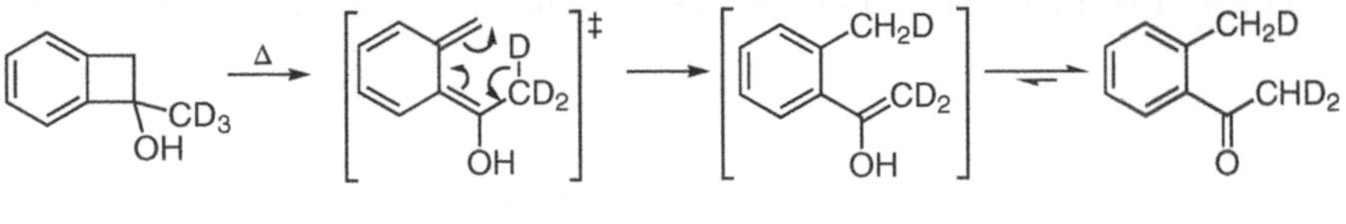

CD$_3$ OH Δ ‡ D CD$_2$ OH CH$_2$D CD$_2$ OH CH$_2$D CHD$_2$ O

11.31. Beaudry, C. M.; Malerich, J. P.; Trauner, D. *Chem. Rev.* **2005**, *105*, 4757; Huisgen, R.; Dahmen, A.; Huber, H. *J. Am. Chem. Soc.* **1967**, *89*, 7130. Both the intermediate and the product are chiral. Only one enantiomer is shown for each.

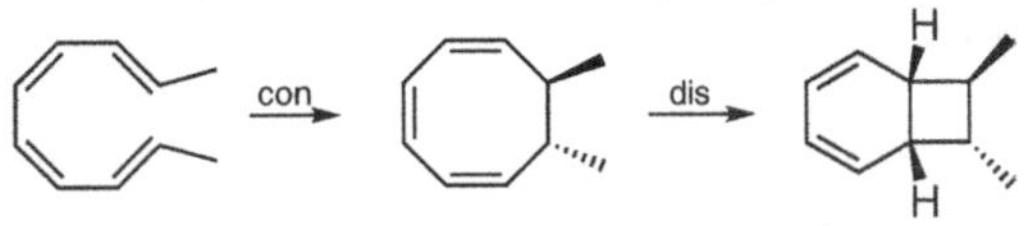

con dis H H

11.32. Wessig, P.; Müller, G. *Chem. Rev.* **2008**, *108*, 2051 and references cited therein. This is a dehydro-Diels–Alder reaction.

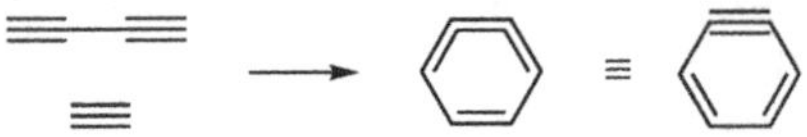

≡

11.33. Houk, K. N.; Strassner, T. *J. Org. Chem.* **1999**, *64*, 800; Wiberg, K. B.; Sagebarth, K. A. *J. Am. Chem. Soc.* **1957**, *79*, 2822. This is a $[_\pi 4_s + _\pi 2_s]$ cycloaddition, followed by hydrolysis of the Mn–O bonds.

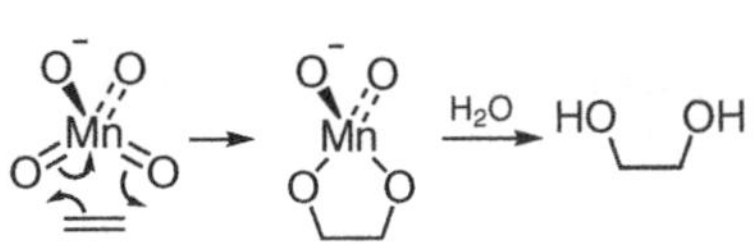

O$^-$ O Mn O O O$^-$ O Mn O O H$_2$O HO OH

11.34. Karmakar, R.; Mamidipalli, P.; Yun, S. Y.; et al. *Org. Lett.* 2013, 15, 1938.

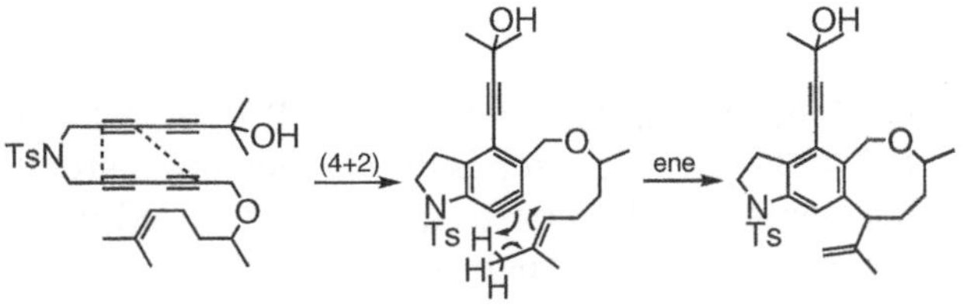

TsN OH O (4+2) OH N Ts O H H H ene OH N Ts O

11.35. Thevenet, N.; de la Sovera, V.; Vila, M. A.; et al. *Org. Lett.* **2015**, *17*, 684.

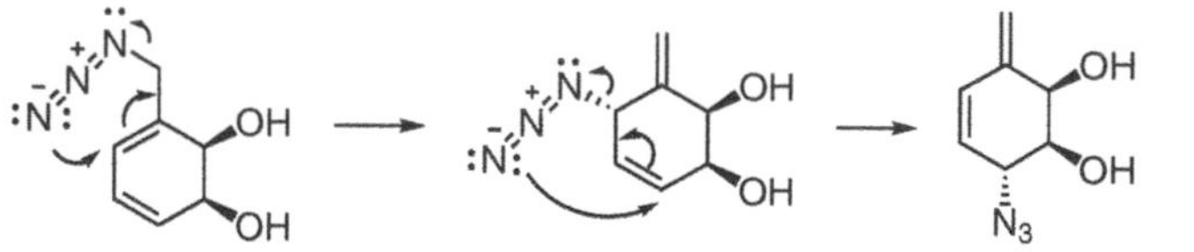

11.36. Frey, H. M.; Walsh, R. *Chem. Rev.* **1969**, 69, 103.

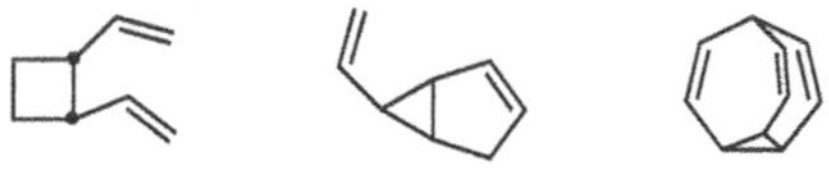

$\Delta S^{\ddagger}$ −11.7 eu −7.2 eu −2.3 eu

11.37. Huisgen, R.; Kalvinisch, I.; Li, X.; et al. *Eur. J. Org. Chem.* **2000**, 1685. See also Morgan, K. M. *Annu. Rep. Prog. Chem. Sect. B* **2001**, *97*, 279. Thiobenzophenone reacts with diazomethane to produce 2,5-dihydro-2,2-diphenyl-1,3,4-thiadiazole. Warming the solution of the adduct to −45 °C releases N_2 and produces the 1,3-dipolar species thiobenzophenone *S*-methylide. Subsequent reaction with another molecule of thiobenzophenone produces the product.

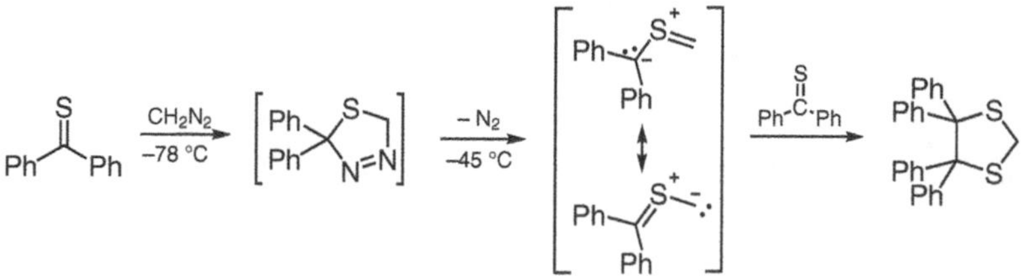

CHAPTER 12

Organic Photochemistry

12.1. From equation 12.1, an absorption onset of 375 nm = 76.3 kcal/mol, and a phosphorescence onset of 6800 Å = 680 nm = 42 kcal/mol. A fluorescence onset of 25,900 cm^{-1} × (0.00286 kcal/mol)/eV = 74.1 kcal/mol.

12.2. They are 3.35 eV (77.3 kcal/mol), 2.61 eV (60.2 kcal/mol), 2.14 eV (49.3 kcal/mol), and 1.79 eV (41.2 kcal/mol), respectively. For hexacene, the S_0-T_1 gap is 12 kcal/mol. Houk, K. N.; Lee, P. S.; Nendel, M. *J. Org. Chem.* **2001**, *66*, 5517.

12.3. **a.** 5×10^{-5} moles of product/10^{-3} Einsteins = 0.05 for Φ_{prod}
b. $1-(0.3+0.5+0.05) = 0.15$

12.4. Compound **O** is anthracene; compound **Q** is DDT. Plotting ln I versus t in each case gives a k value of $2 \times 10^3\ s^{-1}$ for [Q] = 0; k' values are $4.6 \times 10^3\ s^{-1}$ for [Q] = 0.1 M and $5.8 \times 10^3\ s^{-1}$ for [Q] = 0.16 M. Then plotting k'/k (equivalently, τ/τ') versus [Q] produces a Stern–Volmer plot with slope 11.81. Using a value of τ of $1/k$ gives a k_q value of $2.4 \times 10^4\ L\,mol^{-1}\,s^{-1}$.

12.5. Babu, M. K.; Rajasekaran, K.; Kannan, N.; et al. *J. Chem. Soc. Perkin Trans. 2* **1986**, 1721.

$$pK_a^* - pK_a = 8.81, \quad \text{so} \quad pK_a^* = 2.72.$$

12.6. Pacifici, J. G.; Hyatt, J. A. *Mol. Photochem.* **1971**, *3*, 271. This is the ester analogue of a Norrish Type II reaction involving hydrogen abstraction, followed either by reverse hydrogen atom transfer

Solutions Manual for Perspectives on Structure and Mechanism in Organic Chemistry, Third Edition.
Felix A. Carroll.

after bond rotation to form the enantiomer or by fragmentation to the alkene and acid.

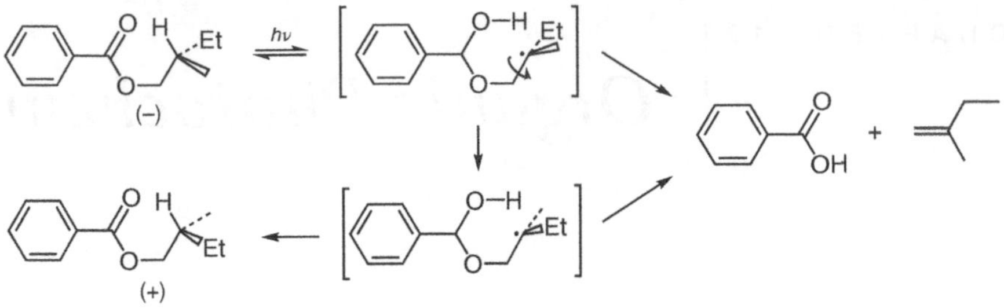

12.7. Frey, H. M.; Lister, D. H. *Mol. Photochem.* **1972**, *3*, 323.

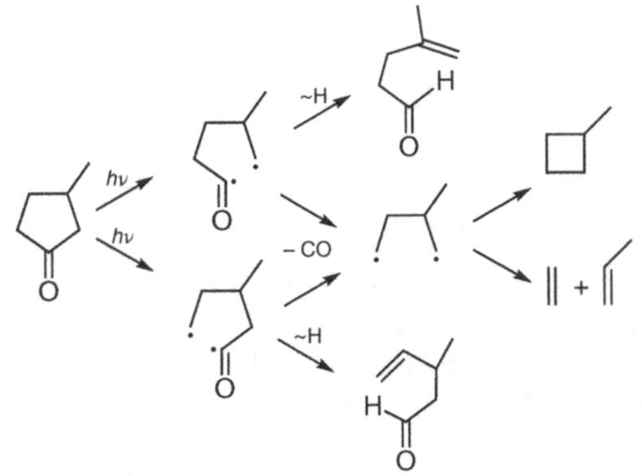

12.8. Tsuneishi, H.; Inoue, Y.; Hakushi, T.; et al. *J. Chem. Soc. Perkin Trans. 2* **1993**, 457. The ratio of isomers present at the photostationary state is $([E]/[Z])_{pss} = 0.23/0.77 = 0.30$. Using the equation

$$([E]/[Z])_{pss} = (k_{dE} / k_{dZ}) \times (\varepsilon_Z / \varepsilon_E)$$

and solving for (k_{dE}/k_{dZ}) yields a value of 0.98. In other words, the excited singlet state formed from either isomer has a nearly equal probability of relaxing to the more stable (*Z*) diastereomer or to the less stable (*E*) diastereomer.

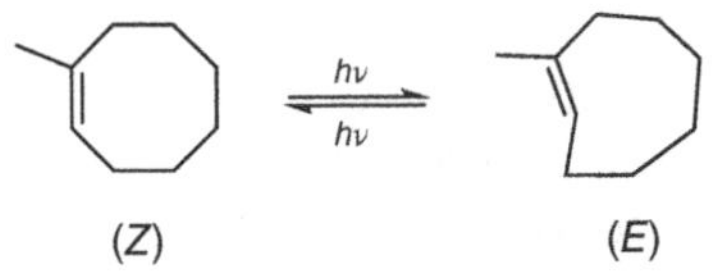

(*Z*) (*E*)

12.9. Alumbaugh, R. L.; Pritchard, G. O.; et al. *J. Phys. Chem.* **1965**, *69*, 3225. α-Cleavage leads to a diradical that can close to the isomer of the reactant, can undergo intramolecular hydrogen atom transfer to form isomeric aldehydes, or can lose CO to form another diradical that can close to isomeric 1,2-dimethylcyclopentanes or undergo intramolecular hydrogen abstraction to form isomeric alkenes.

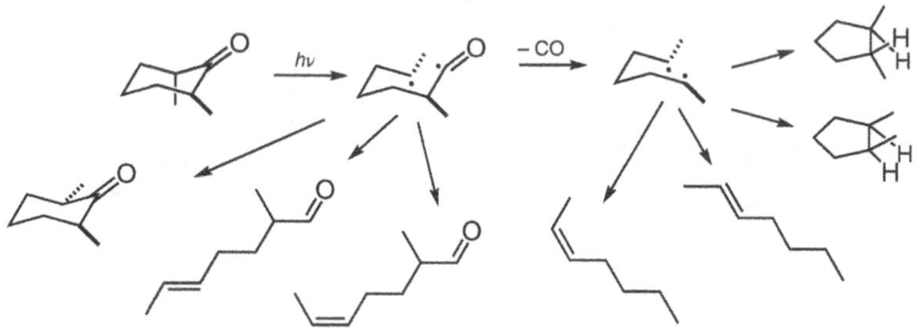

12.10. Ohloff, G.; Klein, E.; Schenck, G. O. *Angew. Chem.* **1961**, *73*, 578. See also Schönberg, A. *Preparative Organic Photochemistry*; Springer-Verlag: New York, 1968; p. 377. The other product is 2,6-dimethyl-1-octene-3,8-diol. It is formed by the same mechanism but with different regiochemistry.

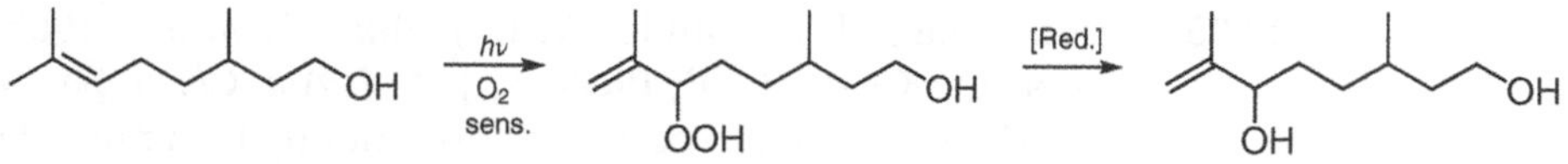

12.11. a. Kropp, P. J. *Mol. Photochem.* 1978–79, *9*, 39 and references therein. In each case, the deuterium is on the carbon adjacent to the carbon bearing the double bond or the methoxy group.

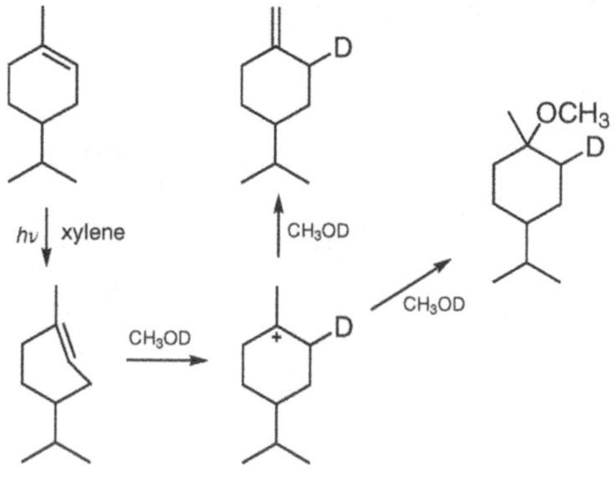

b. Yang, N. C.; Jorgenson, M. J. *Tetrahedron Lett.* **1964**, 1203.

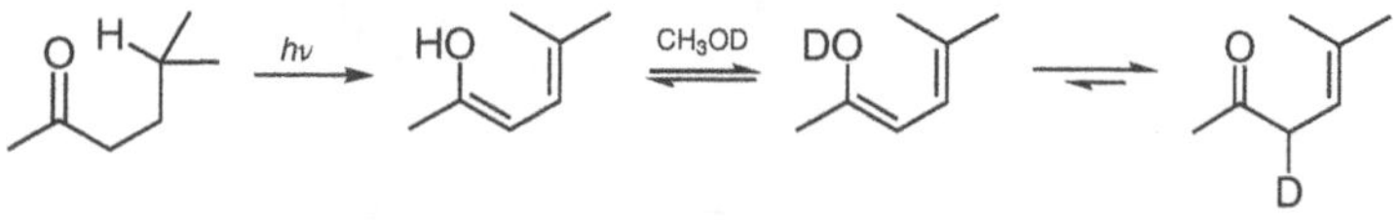

The deuterium label is on C3 due to tautomerization of the dienol-OD formed by equilibration of the dienol-OH with solvent.

12.12. Carless, H. A. J. *J. Chem. Soc. Perkin Trans. 2* **1974**, 834.

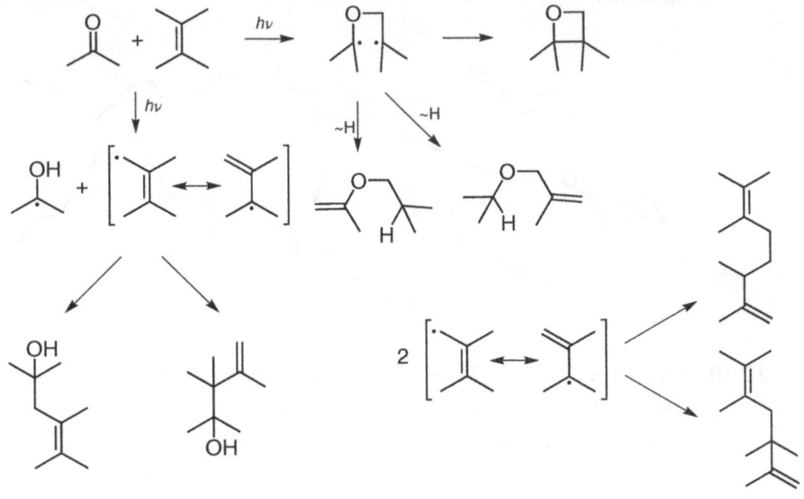

hν

hν

~H

~H

OH

OH

OH

2

12.13. Zimmerman, H. E.; Sandel, V. R. *J. Am. Chem. Soc.* **1963**, *85*, 915. Also see DeCosta, D. P.; Pincock, J. A. *J. Am. Chem. Soc.* **1993**, *115*, 2180. The first reaction occurs predominantly by a radical pathway, while the second is predominantly a photosolvolysis. However, a minor amount of 4-methoxybenzyl alcohol was also formed, perhaps through the intermediacy of a 4-methoxybenzyl cation. There is evidence for some radical reaction in the photochemistry of the meta isomer as well. (Yields shown are the average of values obtained in two experiments.)

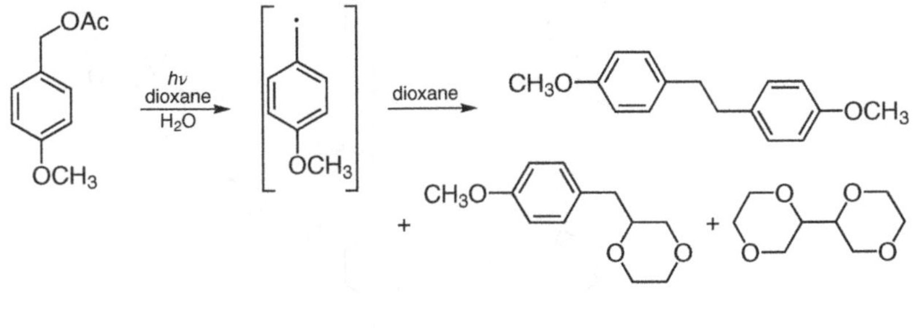

OAc

hν dioxane H_2O

OCH_3

OCH_3

dioxane

CH_3O OCH_3

CH_3O

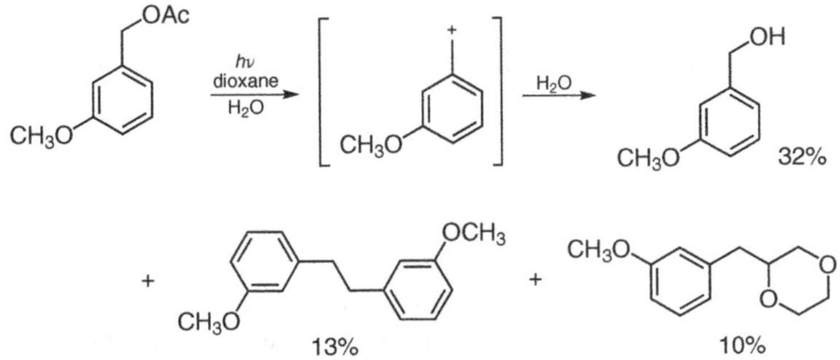

OAc

hν dioxane H_2O

CH_3O

CH_3O

H_2O

OH

CH_3O 32%

OCH_3

CH_3O 13%

CH_3O

10%

12.14. **a.** Butenandt, A.; Poschmann, L. *Chem. Ber.* **1944**, *77B*, 392.

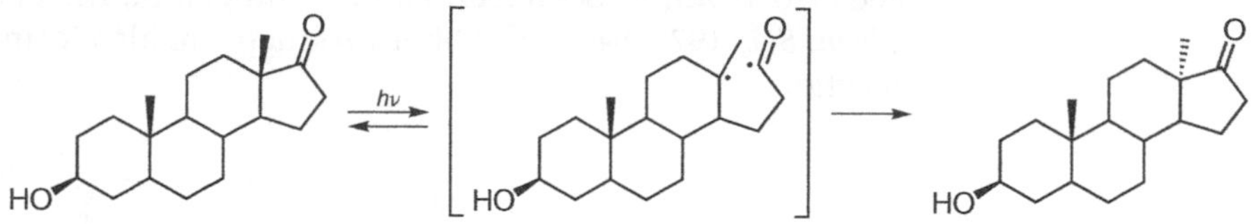

HO hν HO HO

b. Quinkert, G. *Angew. Chem. Int. Ed. Engl.* **1962**, *1*, 166.

HO hν HO ~H HO O=C= H_2O HO CO_2H

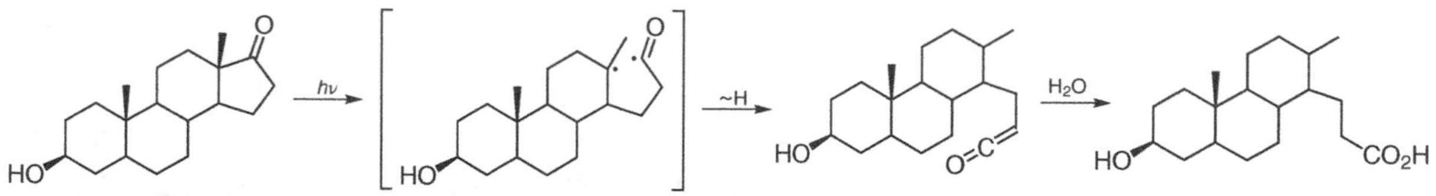

c. Quinkert, G. *Angew. Chem. Int. Ed. Engl.* **1962**, *1*, 166.

O hν O ~H C=C=O H $C_6H_{11}NH_2$ $CONHC_6H_{11}$

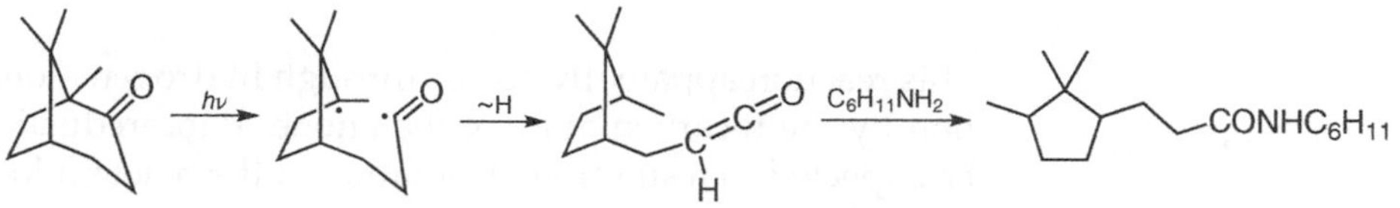

d. Medary, R. T.; Gano, J. E.; Griffin, C. E. *Mol. Photochem.* **1974**, *6*, 107.

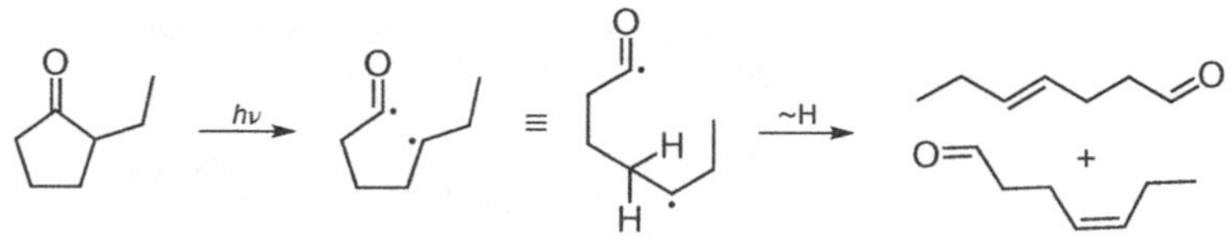

O hν O ≡ O H H ~H O O +

12.15. **a.** Wolff, S.; Schreiber, W. L.; Smith, A. B., III.; et al. *J. Am. Chem. Soc.* **1972**, *94*, 7797. The reaction occurs by abstraction of a hydrogen atom from the methoxy group by the ß-carbon atom of the enone.

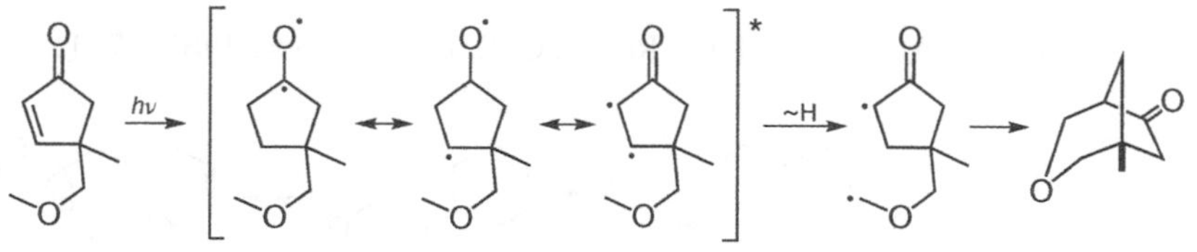

O hν O O O * ~H O O O

b. Barnard, M.; Yang, N. C. *Proc. Chem. Soc.* **1958**, 302.

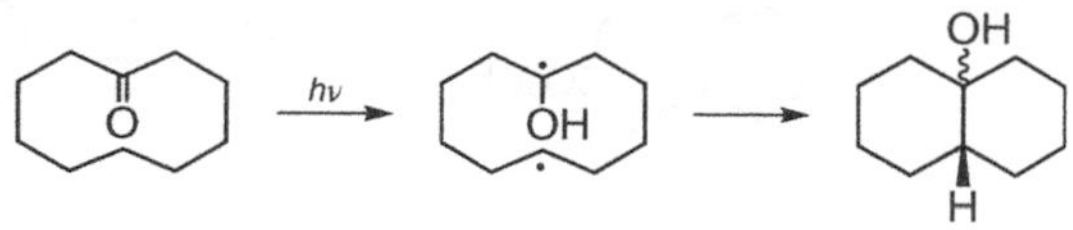

O hν OH OH H

12.16. a. Agosta, W. C.; Smith, A. B., III. *J. Am. Chem. Soc.* **1971**, *93*, 5513. See also Wolff, S.; Schreiber, W. L.; Smith, A. B., III.; et al. *J. Am. Chem. Soc.* **1972**, *94*, 7797. Other products are also formed in the reaction.

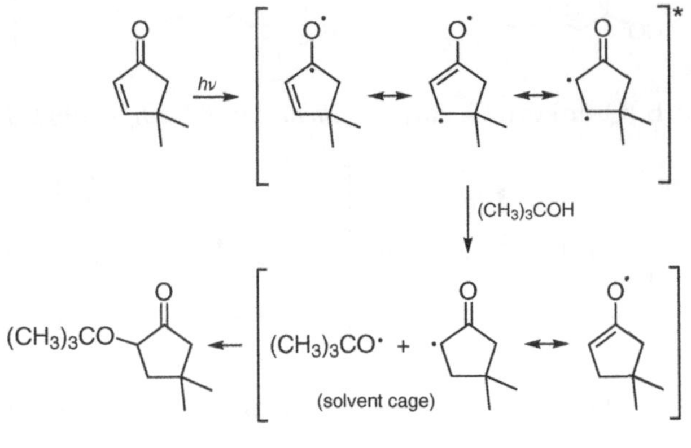
hν
(CH3)3COH
(CH3)3CO
(CH3)3CO· +
(solvent cage)

This reaction apparently occurs through hydrogen atom abstraction by the ß-carbon atom of the enone. Photoreduction would be expected to result from irradiation of the reactant in a 1° or 2° alcohol.

b. Göth, H.; Cerutti, P.; Schmid, H. *Helv. Chem. Acta* **1965**, *48*, 1395.

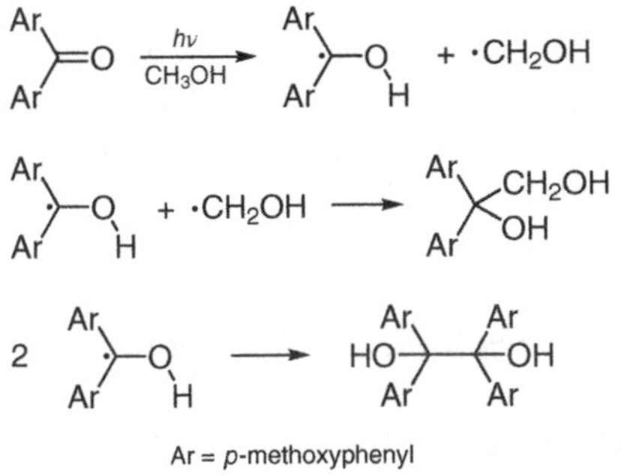
hν
CH3OH
+ ·CH2OH
CH2OH
OH
HO
OH
Ar = p-methoxyphenyl

12.17. a. Anderson, J. C.; Reese, C. B. *J. Chem. Soc.* **1963**, 1781.

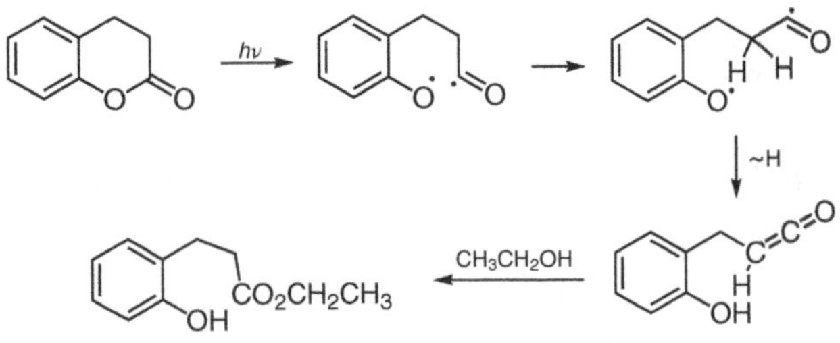
hν
~H
CH3CH2OH
CO2CH2CH3
OH

b. Anderson, J. C.; Reese, C. B. *J. Chem. Soc.* **1963**, 1781.

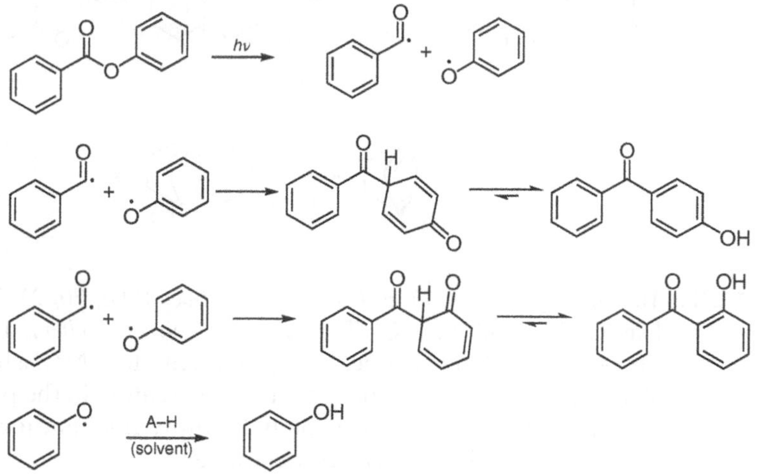

hν

A–H (solvent)

12.18. Padwa, A.; Alexander, E.; Niemcyzk, M. *J. Am. Chem. Soc.* **1969**, *91*, 456. See also Padwa, A. *Acc. Chem. Res.* **1971**, *4*, 48. In isopropyl alcohol solution, intermolecular hydrogen abstraction leads to radicals that combine to give the product observed.

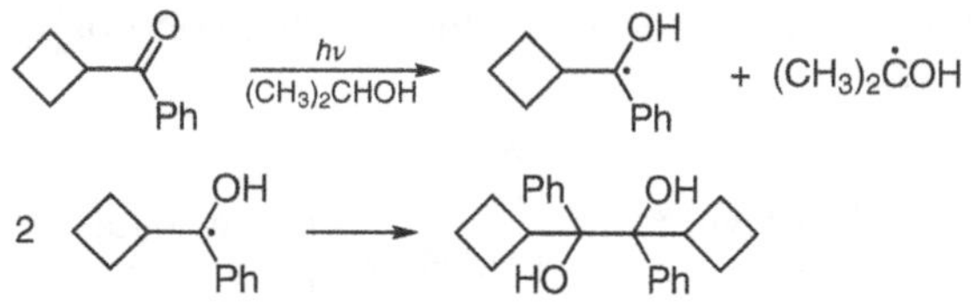

hν, $(CH_3)_2CHOH$; + $(CH_3)_2\dot{C}OH$

In benzene solution, abstraction of hydrogen from solvent is difficult, so intramolecular hydrogen abstraction occurs instead.

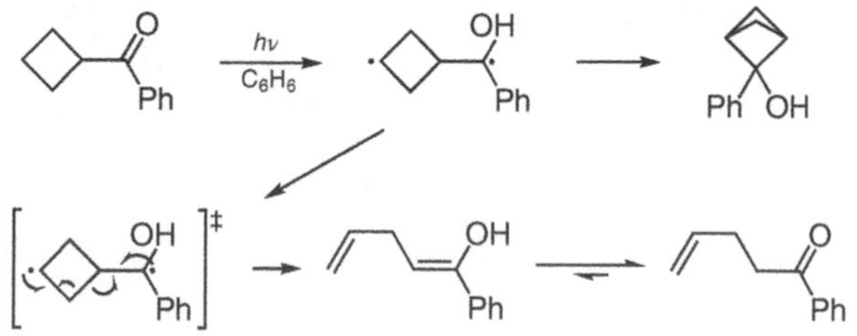

hν, C_6H_6

12.19. a. Bahurel, Y.; Descotes, G.; Pautet, F. *Compt. Rend.* **1970**, *270*, 1528; Bahurel, Y.; Pautet, F.; Descotes, G. *Bull. Soc. Chim. France* **1971**, *6*, 2222.

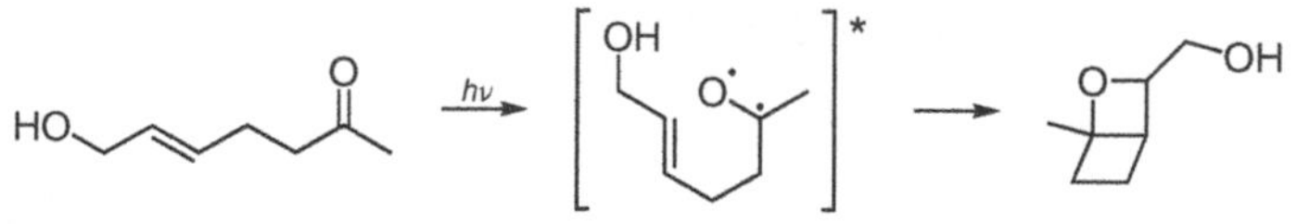

hν

b. Nobs, F.; Burger, U.; Schaffner, K. *Helv. Chim. Acta* **1977**, *60*, 1607. Hydrogen abstraction from the methyl group of the propenyl substituent by the α-carbon atom of the photoexcited enone moiety leads to a diradical that closes to the product in the last step of the reaction.

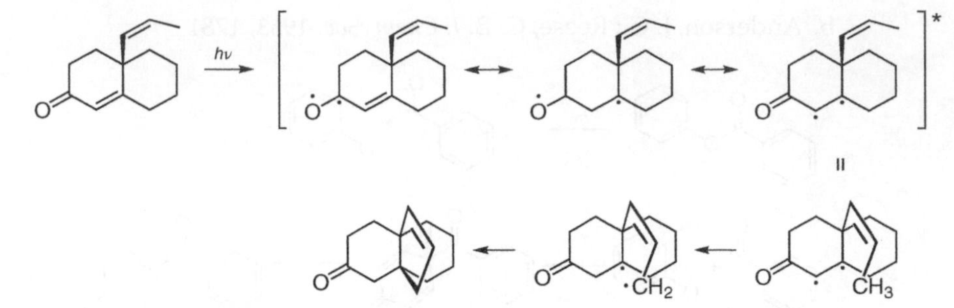

12.20. Turro, N. J.; Weiss, D. S. *J. Am. Chem. Soc.* **1968**, *90*, 2185. See also Dawes, K.; Dalton, J. C.; Turro, N. J. *Mol. Photochem.* **1971**, *3*, 71. There is a stereochemical requirement for Norrish Type II abstraction. Specifically, only a γ-hydrogen atom in the plane of the carbonyl group (that is near the half-empty nonbonding orbital of the n,π^* excited ketone) can be abstracted.

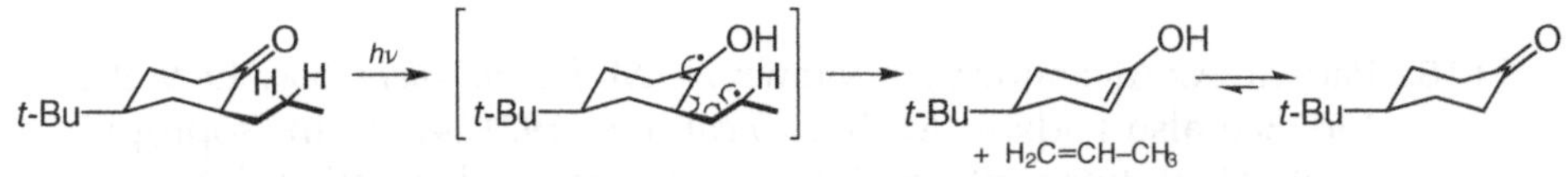

In the absence of a γ-hydrogen atom in this position, only Norrish Type I cleavage occurs, which leads to photoepimerization.

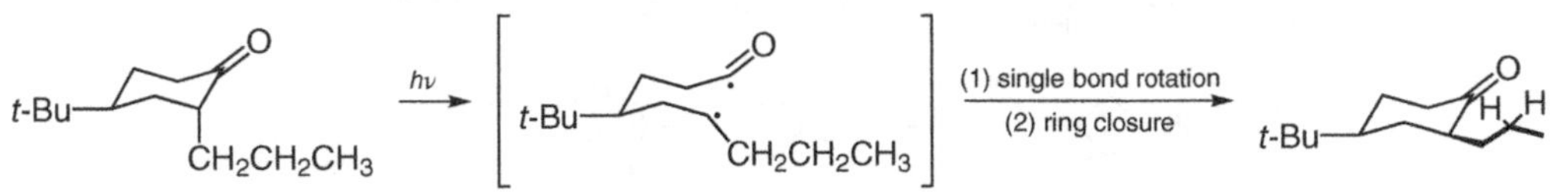

12.21. a. Padwa, A.; Eisenberg, W. *J. Am. Chem. Soc.* **1970**, *92*, 2590.

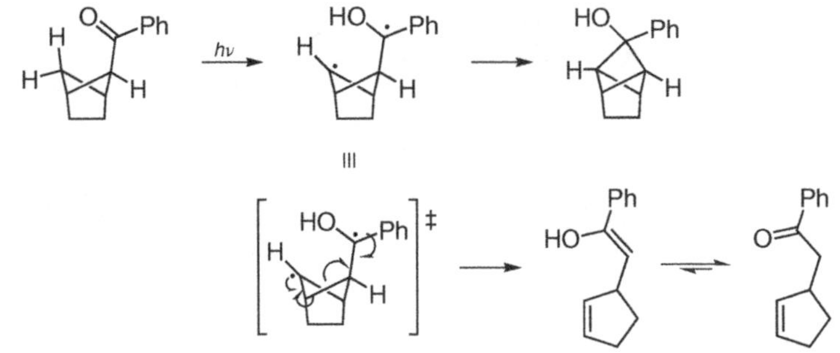

b. Gagosian, R. B.; Dalton, J. C.; Turro, N. J. *J. Am. Chem. Soc.* **1970**, *92*, 4752; see also Turro, N. J.; Dalton, J. C.; Dawes, K.; et al. *Acc. Chem. Res.* **1972**, *5*, 92.

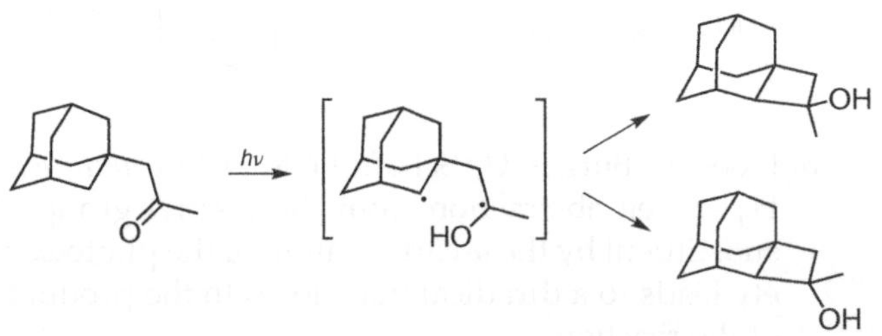

c. Padwa, A.; Alexander, E.; Niemcyzk, M. *J. Am. Chem. Soc.* **1969**, *91*, 456; see also Padwa, A. *Acc. Chem. Res.* **1971**, *4*, 48.

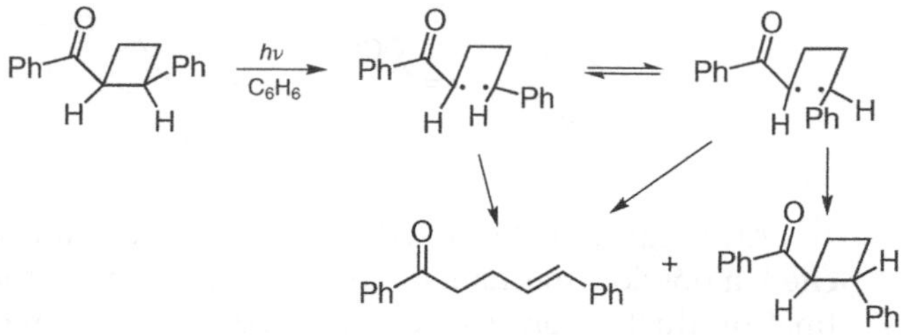

hν
C_6H_6

d. Arnold, B. J.; Mellows, S. M.; Sammes, P. G.; et al. *J. Chem. Soc. Perkin Trans. 1* **1974**, 401. See also Durst, T.; Kozma, E. C.; Charlton, J. L. *J. Org. Chem.* **1985**, *50*, 4829.

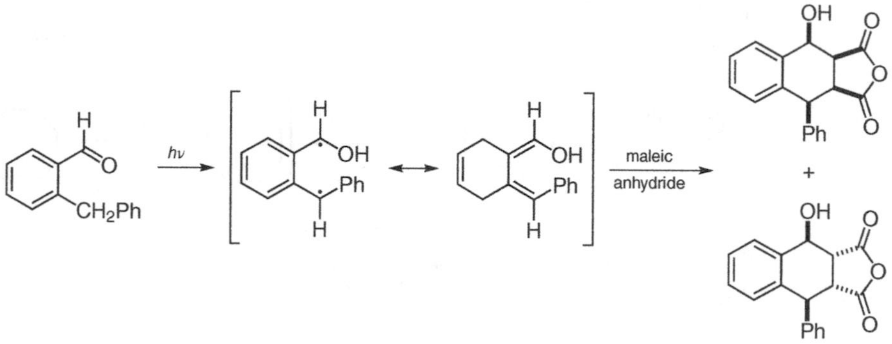

hν
maleic anhydride

e. Wolff, S.; Schreiber, W. L.; Smith, A. B., III.; et al. *J. Am. Chem. Soc.* **1972**, *94*, 7797. Abstraction of the 3° hydrogen from the 3-methylbutyl substituent by the ß carbon atom of the enone leads to a biradical. Disproportionation by hydrogen abstraction from a carbon adjacent to the radical center on the alkyl chain leads to the first two products; closure of the diradical leads to the third.

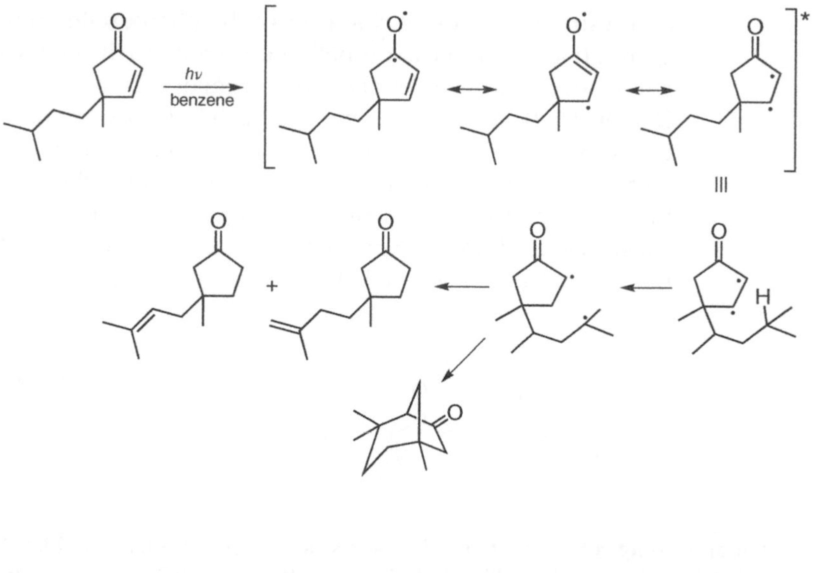

hν
benzene

12.22. Gilbert, A.; Taylor, G. N.; bin Samsudin, M. W. *J. Chem. Soc. Perkin Trans. 1* **1980**, 869. The intermediate is a 1,2-photoadduct.

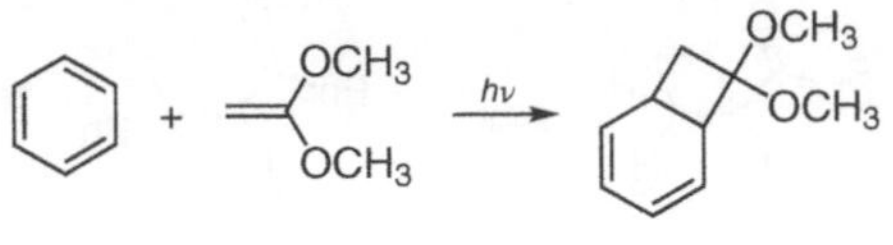

Although not discussed explicitly in the literature reference cited, a possible mechanism for the conversion of the adduct to the final product involves acid-catalyzed loss of methanol, rearrangement (ring opening) of the carbocation, and methyl group displacement.

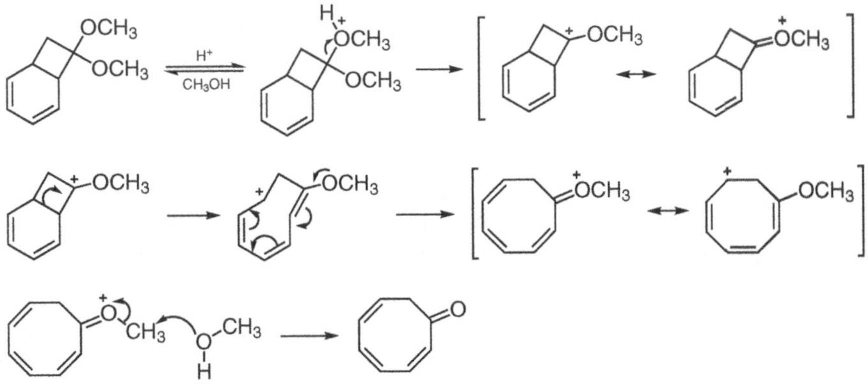

12.23. Morrison, H.; Pajak, J.; Peiffer, R. *J. Am. Chem. Soc.* **1971**, *93*, 3978. The data suggest that the benzene ring absorbs the excitation, undergoes intersystem crossing to the aromatic triplet, and transfers energy to the olefin (intramolecular triplet sensitization). The olefinic triplet then decays to the cis and trans isomers of the starting material.

12.24. Photochemical reaction of *o*-nitrobenzyl compounds is thought to occur through abstraction of a hydrogen atom from the benzyl group as a result of the radical character of the photoexcited nitrobenzene group. If an ether or ester function is bonded to the benzylic carbon, *o*-nitrosobenzaldehhydes or *o*-nitrosobenzophenones can be formed.[1] Riguet, E.; Bochet, C. G. *Org. Lett.* **2007**, *9*, 5453 reported that an unprotonated amine group para to the nitro function results in intramolecular charge transfer, thus preventing the photoexcited nitro group from abstracting hydrogen from the ortho benzylic substituent. Adding HCl protonates the amine nitrogen and prevents the charge transfer shown below from occurring.

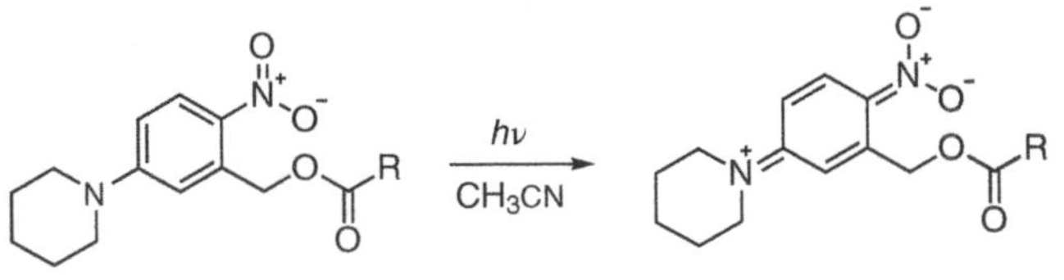

[1] For an leading references to mechanistic studies, see Schmierer, T.; Bley, F.; Schaper, K.; Gilch, P. *J. Photochem. Photobiol. A Chem.* **2011**, *217*, 363; Zhao, H.; Sterner, E. S.; Coughlin, E. B.; Theatro, P. *Macromolecules* **2012**, *45*, 1723.

12.25. The products are the cis,trans, cis,cis, and trans,cis isomers. Kozukue, N.; Park, M.-S.; Choi, S.-H.; et al. *J. Agric. Food Chem.* **2007**, *55*, 7131.

12.26. a. γ-hydrogen abstraction from the triplet state of the phenone moiety leads to a diradical that relaxes to an enol, and in turn it undergoes a [$_4\pi$] conrotatory closure to the cyclobutanol. See Moorthy, J. N.; Samanta, S. *ARKIVOC* **2007** (viii), 324.

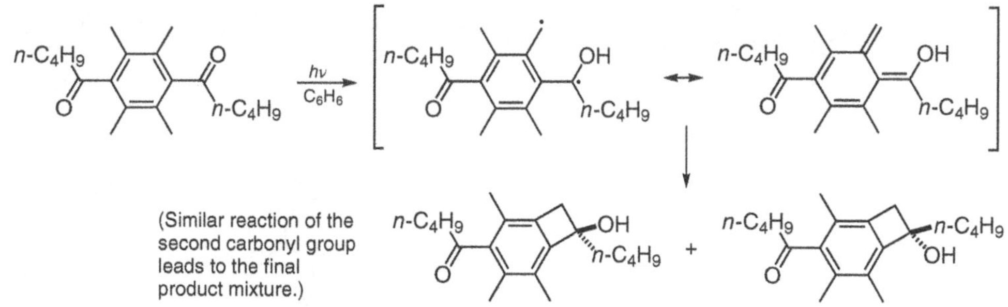

(Similar reaction of the second carbonyl group leads to the final product mixture.)

b. Moorthy, J. N.; Samanta, S. *ARKIVOC* **2007**, 324.

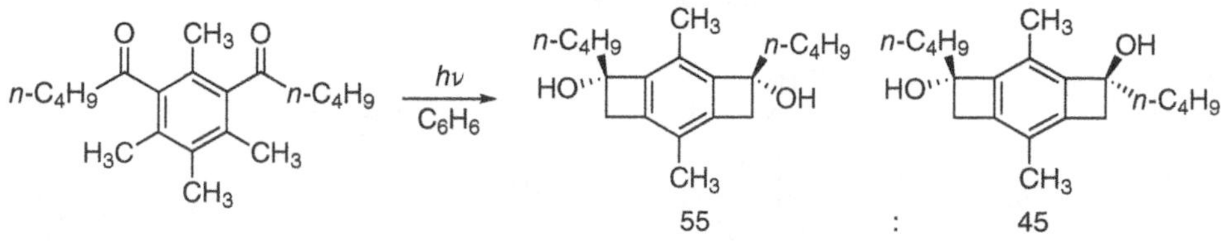

12.27. a. Turro, N. J.; Lee, T.-J. *J. Am. Chem. Soc.* **1969**, *91*, 5651.

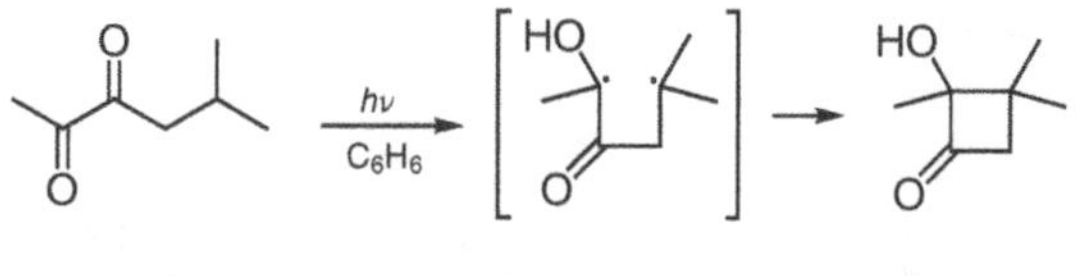

b. Corey, E. J.; Nozoe, S. *J. Am. Chem. Soc.* **1964**, *86*, 1652.

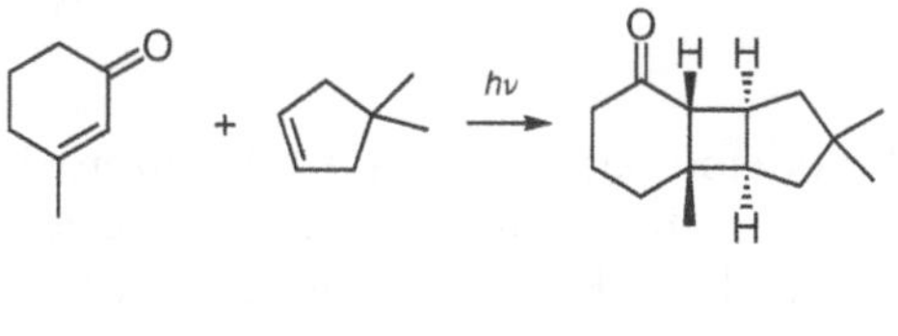

12.28. Crimmins, M. T.; Pace, J. M.; Nantermet, P. G.; et al. *J. Am. Chem. Soc.* **2000**, *122*, 8453.

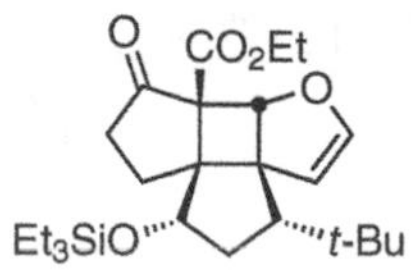

12.29. Kang, T.; Scheffer, J. R. *Org. Lett.* **2001**, *3*, 3361. Norrish Type I fragmentation followed by hydrogen abstraction within the crystal leads to cyclopentene and benzaldehyde, which then undergo oxetane formation.

12.30. a. Winkler, J. D.; Lee, E. C. Y. *J. Am. Chem. Soc.* **2006**, *128*, 9040.

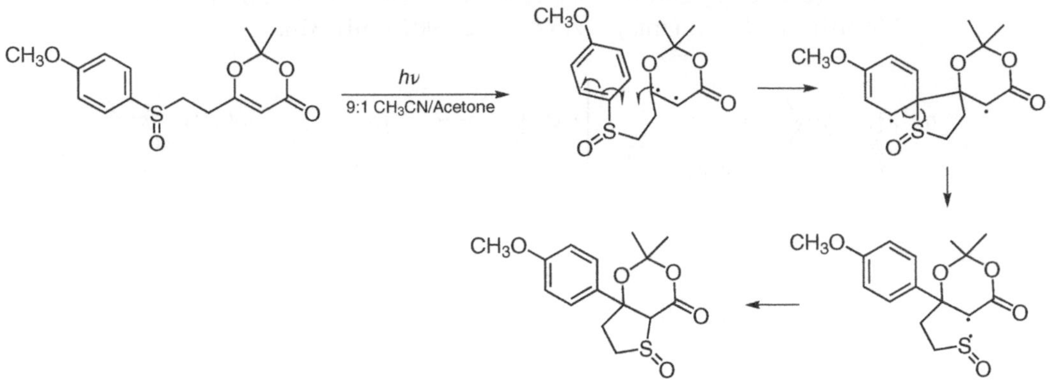

CH_3O hν 9:1 CH_3CN/Acetone

b. Sajimon, M. C.; Ramaiah, D.; Suresh, C. H.; et al. *J. Am. Chem. Soc.* **2007**, *129*, 9439.

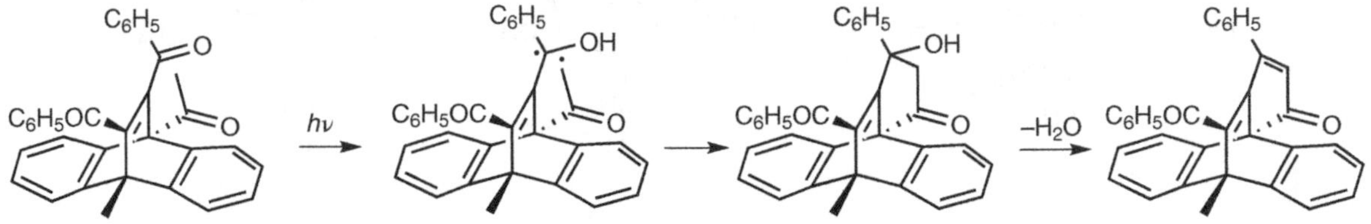

C_6H_5 C_6H_5OC hν OH $-H_2O$

c. Chang, J. A.; Chiang, Y.; Keeffe, J. R.; et al. *J. Org. Chem.* **2006**, *71*, 4460.

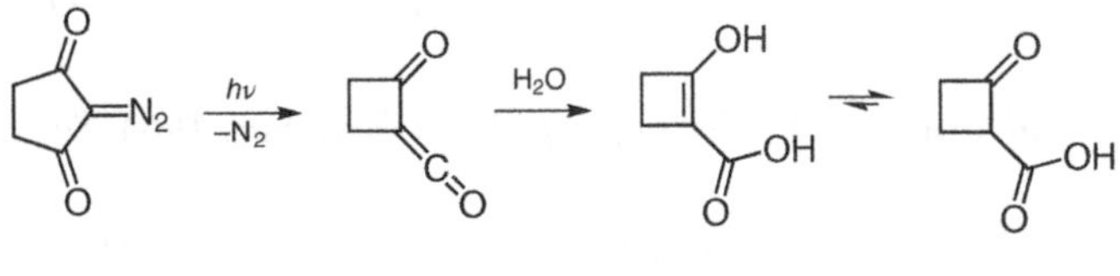

$=N_2$ hν $-N_2$ H_2O OH

d. Photodeiodination gives a tetrafluoro-1,4-benzyne that undergoes retro-Bergman cyclization. See reference 31 in Klein, M.; Walenzyk, T.; König, B. *Collect. Czech. Chem. Commun.* **2004**, *59*, 945.

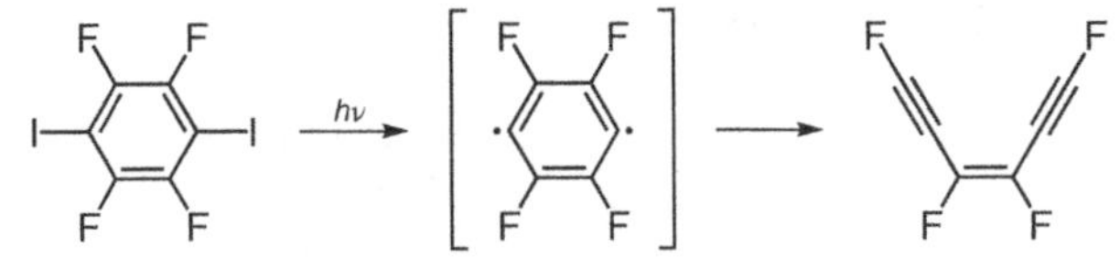

F I hν

12.31. Shibamoto, T. *J. Toxicol. Cut. Ocular Toxicol.* 1983–84, *2*, 267. α-Cleavage leads to benzyl and acyl radicals that can recombine to produce the methylacetophenones after tautomerization. Coupling of two benzyl radicals leads to bibenzyl. Abstraction of hydrogen from a hydrogen donor by benzyl radical leads to toluene.

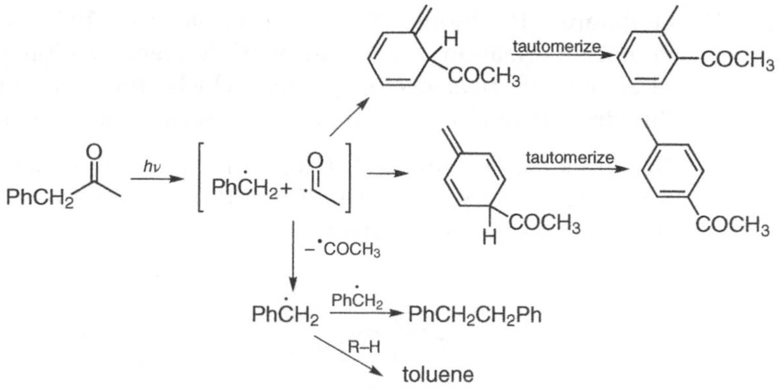

12.32. This is a di-π-methane rearrangement. Hixson, S. S.; Mariano, P. S.; Zimmerman, H. E. *Chem. Rev.* **1973**, *73*, 531.

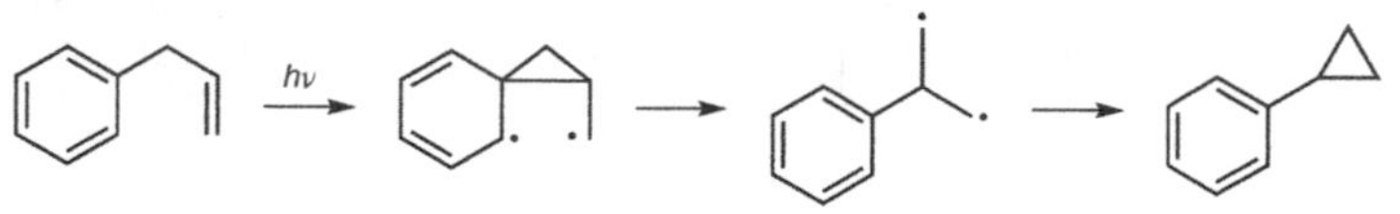

12.33. Tedaldi, L. M.; Baker, J. R. *Org. Lett.* **2009**, *11*, 811. The intramolecular Paternò–Büchi reaction leads to the structure in brackets. In the absence of a reducing agent, the carbonyl group absorbs UV light and undergoes further reaction, thus leaving only a small amount after irradiation ends. With $LiBH_4$ present, however, the carbonyl group is reduced to the alcohol and further photoreaction is prevented.

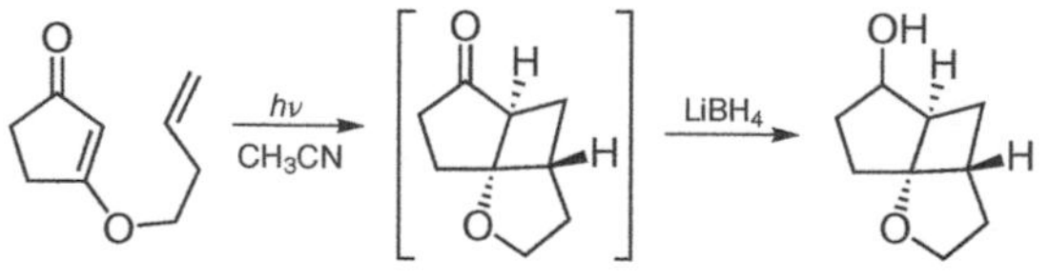

12.34. Hamer, N. K. *J. Chem. Soc. Perkin 1* **1979**, 1285. The first step is a hydrogen abstraction and ring closure to produce 2-hydroxy-2-methyl-2,3-dihydro-1H-inden-1-one. It then undergoes α-cleavage to produce the final product after transfer of the hydroxyl hydrogen to the acyl radical center.

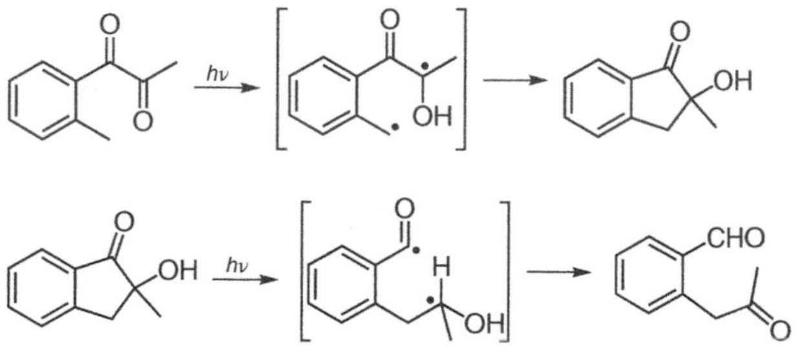

12.35. Hoffmann, R.; Inoue, Y. *J. Am. Chem. Soc.* **1999**, *121*, 10702. Photoisomerization as a result of triplet sensitization produces a racemate of *trans*-cycloheptene, which then undergoes syn dihydroxylation by OsO_4 to produce the racemate of the 1,2-diol.

12.36. Griesbeck, A. G.; Hinze, O.; Görner, H.; et al. *Photochem. Photobiol. Sci.* **2012**, *11*, 587. Norrish Type II photofragmentation, with acetophenone as a byproduct.

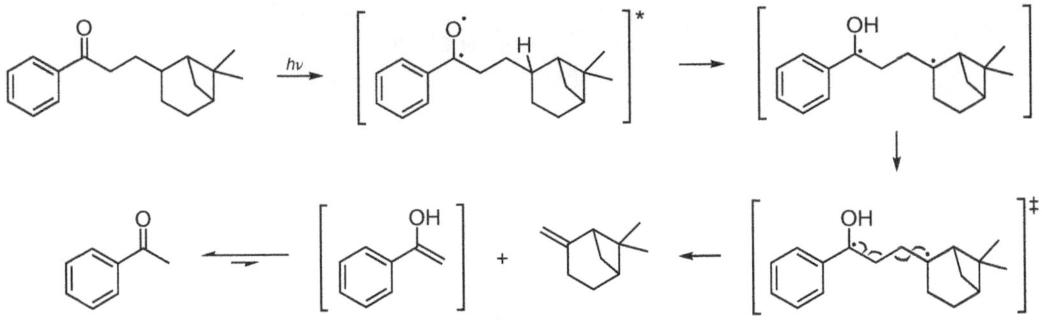